GW00722661

Integral Equation Methods in
Potential Theory and Elastostatics

COMPUTATIONAL MATHEMATICS AND APPLICATIONS

Series Editor

J. R. WHITEMAN

Institute of Computational Mathematics, Brunel University, England.

E. HINTON and D. R. J. OWEN: Finite Element Programming

M. A. JASWON and G. T. SYMM: Integral Equation Methods in Potential Theory and Elastostatics

Integral Equation Methods in Potential Theory and Elastostatics

1977

M. A. JASWON

Professor of Mathematics and Head of Department
The City University, London

and

G. T. SYMM

Division of Numerical Analysis and Computing
National Physical Laboratory, Teddington

1977

ACADEMIC PRESS
London   New York   San Francisco
A Subsidiary of Harcourt Brace Jovanovich, Publishers

ACADEMIC PRESS INC. (LONDON) LTD.
24/28 Oval Road,
London NW1

United States Edition published by
ACADEMIC PRESS INC.
111 Fifth Avenue
New York, New York 10003

Library of Congress Catalog Card Number: 77-74375

ISBN: 0-12-381050-7

Printed in Gt Britain by Page Bros (Norwich) Ltd., Norwich

Editor's Foreword

This new series on computational mathematics has been conceived in order to fill the gap which exists between numerical mathematics and theoretical applied mathematics on the one hand and engineering and scientific applications of numerical methods on the other. The numerical theme is central to the series and thus topics suitable for inclusion will range from numerical analysis to the application of numerical methods in engineering. The series will consist of text books, monographs and conference proceedings spanning computational mathematics and its applications. Naturally some books will be so specialised as to be limited either to theory or to practice. However, the emphasis throughout will be on a readable presentation and it is intended that all the books in the series will be tools from which persons can readily learn.

The present volume fits perfectly into the general scheme of the series in that Professor Jaswon and Dr. Symm have brought together in one book theory and practice. The first part of this volume gives a very clear description of the theory of integral equations in potential theory and elastostatics. In particular it provides the first unifying account of scalar and vector potential theory and includes much previously unpublished work. In the second part numerical methods for solving the integral equation formulations of practical problems, based on the authors' extensive research, are discussed. The book is so written that parts 1 and 2 can be read more or less independently, part 2 needing only occasional reference back to part 1. This latter facet will prove invaluable to persons interested primarily in numerical methods.

Brunel University   J. R. Whiteman
July 1977

v

Preface

Integral equation methods are becoming increasingly accepted as a means for solving the boundary-value problems of potential theory and elastostatics. They often provide effective solutions which fall beyond the scope of other approaches. This monograph provides the first unified account of scalar and vector potential theory using a common formalism, and it includes much previously unpublished material. The theory is illustrated by a wealth of numerical solutions to problems drawn from a variety of fields, including electrostatics, potential fluid flow, heat conduction, and the stretching and bending of thin plates.

In order to meet the varying requirements of readers, the book has been divided into two fairly distinct parts. Part I concentrates upon the theory while Part II gives a largely self-contained account of the applications. Readers who are interested primarily in numerical aspects may proceed immediately to Part II, where necessary back references to Part I are clearly indicated. An extensive list of references is provided, which also covers recent interesting numerical solutions of problems in three-dimensional elastostatics.

The book should be of direct use to engineers interested in a new approach to any of the fields mentioned above, and also to postgraduate students who require a guide to this rapidly evolving subject. The theoretical material has served as the basis of an M.Sc. course at The City University.

We should like to acknowledge the benefit of helpful advice from Dr. I. C. Brown, Mr. A. J. Burton, Dr. J. Cassells, Professor R. E. Gibson, Dr. A. W. Gillies, Mr. M. G. Hassan, Mr. G. F. Miller, Dr. M. G. M. Smith, Mr. R. J. Smith, Dr. T. E. Stanley, Professor J. R. Whiteman and Professor J. R. Willis, some of whom are mentioned at appropriate points in the text. We should also like to acknowledge the provision of facilities at The City University and the National Physical Laboratory. Finally, our grateful thanks are due to Mr. R. A. Farrand, Academic Press, Ltd., for his patience and ready cooperation.

July 1977
M.A.J.
G.T.S.

Contents

PART I: Theory

1 Elements of Potential Theory

2 Representation of Harmonic Functions by Potentials

3 Green's Formula

PART II: Applications

Introduction

Integral equations provide a distinctive formulation of the fundamental boundary-value problems of potential theory. The existence of solutions to such equations was first rigorously demonstrated by Fredholm (1903) on the basis of his celebrated discretisation procedure. It was not envisaged by Fredholm or his immediate successors that solutions could actually be constructed in this way. However, the advent of fast digital computers (circa 1960) has made it possible to implement discretisation processes arithmetically, so enabling numerical solutions of tolerable accuracy to be readily achieved. These solutions define source densities over the relevant boundary, which generate approximate values of the required potential field at any selected set of interior or exterior points as the case may be. No important limitations, either on the shape of the boundary or on the connectivity of the domain which it encloses, are necessary. This developing theory and its applications form the basis of the present book.

Fredholm integral equations follow from the representation of harmonic functions by simple-layer or double-layer potentials. An alternative approach is through Green's formula, which represents a harmonic function as the superposition of a simple-layer and a double-layer potential. Taking the field point on the boundary, we obtain a functional constraint between the boundary values and normal derivatives of the harmonic function which yields integral equation formulations. These are related to the Fredholm formulations but receive no mention, for instance, in Kellogg's (1929) classical treatise. Because of this and their importance from the numerical viewpoint, we accord them some emphasis at the expense of better known material.

Vector potentials analogous to the scalar simple-layer and double-layer potentials were first introduced by Kupradze (1965). These provide a representation of linear elastostatic displacement fields analogous to the representation of harmonic functions by scalar potentials. As a consequence, the fundamental boundary-value problems of linear elastostatics may be formulated by vector integral equations analogous to the Fredholm integral equations of potential theory. There has been considerable progress since

1965, particularly in the exploitation of Somigliana's formula. When the field point lies on the boundary we obtain a vector functional constraint between boundary displacements and tractions, which yields integral equation formulations allied to the Kupradze formulations. Somigliana's boundary formula is of course the vector analogue of Green's boundary formula, displacement and traction corresponding respectively with the harmonic function and its normal derivative. A suggestive symbolism has been adopted in this book, which enables the vector and scalar theories to be immediately correlated at each stage.

The elastic continuum may be either isotropic or anisotropic. Our foregoing remarks apply to each though, as might be expected, the isotropic theory has been carried much further. Linear problems of the stretching and bending of isotropic plates are governed by the biharmonic equation, which leads to specialised representations and boundary formulations. A systematic account of these, including the role played by multi-valued harmonic and biharmonic functions, is presented here for the first time.

Part I of this book gives a treatment of scalar and vector potential theory directed towards the formulation of boundary integral equations. Part II describes an elementary numerical treatment of these equations with illustrative examples, drawing upon published material, Ph.D. theses and special reports. Some of the problems solved are of direct technological interest in engineering fields. Unfortunately no rigorous error analysis of the numerical solutions is available. However, various computer-dependent procedures have been introduced, which enable the accuracy of the solutions to be assessed a posteriori. These procedures include an examination of numerical conditioning as the number of pivotal points increases; the solution of suitable trial problems for the domain under consideration; and appeals to maximum principles where appropriate. To summarize, a new area of applied numerical analysis has been opened up in response to the potentialities inherent in modern digital computers.

Part I

Theory

1

Elements of Potential Theory

1.1 Discrete distribution of simple sources

In this chapter we collect the main facts of classical potential theory, as found for instance in Kellogg (1929), and we write them out in a form that will be found convenient for subsequent applications.

A unit simple source, located at a point $\mathbf{q}$ referred to a suitable three-dimensional coordinate system, generates the Newtonian potential

$$g(\mathbf{p}, \mathbf{q}) = g(\mathbf{q}, \mathbf{p}) = |\mathbf{p} - \mathbf{q}|^{-1} \qquad (1.1.1)$$

at a field point $\mathbf{p}$. This potential is a continuous function of $\mathbf{p}$, differentiable to all orders, and it satisfies Laplace's equation

$$\nabla^2 g(\mathbf{p}, \mathbf{q}) = 0$$

everywhere except at the source point $\mathbf{q}$. These properties characterise g as a harmonic function of $\mathbf{p}$ everywhere except at $\mathbf{p} = \mathbf{q}$. Formally g satisfies Poisson's equation

$$\nabla^2 g(\mathbf{p}, \mathbf{q}) = -4\pi\delta(\mathbf{p} - \mathbf{q}) \qquad (1.1.2)$$

everywhere, where δ is Dirac's delta function centred upon $\mathbf{q}$ (Dirac, 1935; Lighthill, 1958; Jones, 1966). The use of the delta function provides a quick approach to results which could otherwise be proved only by limiting processes of classical analysis.

A discrete distribution of simple sources, of strengths $\sigma_1, \sigma_2, \ldots, \sigma_N$

located at $\mathbf{q}_1, \mathbf{q}_2, \ldots, \mathbf{q}_N$ respectively, generates the Newtonian potential

$$V(\mathbf{p}) = \sum_{\alpha=1}^{N} g(\mathbf{p}, \mathbf{q}_\alpha)\sigma_\alpha \qquad (1.1.3)$$

at $\mathbf{p}$. This has essentially the same properties as $g(\mathbf{p}, \mathbf{q})$; in particular it is a harmonic function everywhere except at $\mathbf{q}_1, \mathbf{q}_2, \ldots, \mathbf{q}_N$ and it formally satisfies Poisson's equation

$$\nabla^2 V(\mathbf{p}) = -4\pi \sum_{\alpha=1}^{N} \delta(\mathbf{p} - \mathbf{q}_\alpha)\sigma_\alpha \qquad (1.1.4)$$

everywhere. If the source points are located within a finite region of space, then

$$g(\mathbf{p}, \mathbf{q}_\alpha) = |\mathbf{p}|^{-1} + |\mathbf{p}|^{-3}(\mathbf{p} \cdot \mathbf{q}_\alpha) + O(|\mathbf{p}|^{-3}) \quad \text{as} \quad |\mathbf{p}| \to \infty \qquad (1.1.5)$$

and it follows that

$$V(\mathbf{p}) = |\mathbf{p}|^{-1} \sum_{\alpha=1}^{N} \sigma_\alpha + |\mathbf{p}|^{-3} \sum_{\alpha=1}^{N} (\mathbf{p} \cdot \mathbf{q}_\alpha)\sigma_\alpha + O(|\mathbf{p}|^{-3}) \quad \text{as} \quad |\mathbf{p}| \to \infty \qquad (1.1.6)$$

Integrating both sides of (1.1.4) through a domain B bounded by a smooth surface ∂B which encloses all the sources, we find $\partial = boundary\ operator$

$$\int_B \nabla^2 V(\mathbf{p})\,dP = -4\pi \sum_{\alpha=1}^{N} \sigma_\alpha \int_B \delta(\mathbf{p} - \mathbf{q}_\alpha)\,dP = -4\pi \sum_{\alpha=1}^{N} \sigma_\alpha,$$

where dP denotes the volume element at $\mathbf{p}$. But

$$\int_B \nabla^2 V(\mathbf{p})\,dP = \int_{\partial B} \frac{\partial V(\mathbf{p})}{\partial n}\,dp \qquad (1.1.7)$$

by a formal application of the divergence theorem, where $\partial V(\mathbf{p})/\partial n$ denotes the outward normal derivative of V at $\mathbf{p} \in \partial B$ and dp denotes the surface area element at $\mathbf{p} \in \partial B$. Hence

$$\int_{\partial B} \frac{\partial V(\mathbf{p})}{\partial n}\,dp = -4\pi \sum_{\alpha=1}^{N} \sigma_\alpha. \qquad (1.1.8)$$

This is the Gauss flux theorem, which may be verified directly for a large spherical bounding surface by an application of (1.1.6). If ∂B does not enclose any of the sources, i.e. if V is a harmonic function throughout B, it follows that

$$\int_{\partial B} \frac{\partial V(\mathbf{p})}{\partial n} \, dp = 0. \tag{1.1.9}$$

1.2 Volume distribution of simple sources

We now introduce a continuous distribution of simple sources of volume density $\sigma(\mathbf{q})$ at $\mathbf{q}$ extending through a domain B assumed to be regular in the sense defined by Kellogg (see Appendix 1). This generates the potential

$$V(\mathbf{p}) = \int_B g(\mathbf{p}, \mathbf{q})\sigma(\mathbf{q}) \, dQ \tag{1.2.1}$$

which is continuous and differentiable to the first order everywhere. It is differentiable to the second order, and it satisfies Poisson's equation

$$\nabla^2 V(\mathbf{p}) = -4\pi \int_B \delta(\mathbf{p} - \mathbf{q})\sigma(\mathbf{q}) \, dQ = -4\pi\sigma(\mathbf{p}) \tag{1.2.2}$$

at any point $\mathbf{p} \in B$ provided that σ satisfies a Hölder condition at $\mathbf{p}$. Roughly speaking this condition is stronger than continuity but weaker than differentiability (see Appendix 2). Since $\sigma = 0$ outside B, V is a harmonic function everywhere outside B with asymptotic behaviour

$$V(\mathbf{p}) = |\mathbf{p}|^{-1} \int_B \sigma(\mathbf{q}) \, dQ + O(|\mathbf{p}|^{-2}) \quad \text{as} \quad |\mathbf{p}| \to \infty, \tag{1.2.3}$$

where only the first term is radially symmetric.

As a simple exercise suppose B to be a sphere of radius a, with $\sigma(\mathbf{p}) = \sigma_0$ (constant) throughout B, in which case V is radially symmetric everywhere. Hence, by virtue of (1.2.3),

$$V(\mathbf{p}) = |\mathbf{p}|^{-1} \int_B \sigma(\mathbf{q}) \, dQ; \qquad \mathbf{p} \text{ outside } B,$$

$$\text{i.e. } V = r^{-1} \frac{4\pi a^3}{3} \sigma_0; \qquad |\mathbf{p}| = r \geqslant a, \tag{1.2.4}$$

using a more familiar notation. Within B, since $\nabla^2 r^2 = 6$, V must by virtue of (1.2.2) be of the form

$$V = c - \frac{2\pi r^2}{3}\sigma_0; \qquad r \leqslant a, \tag{1.2.5}$$

where c is a constant as yet undetermined. Equating (1.2.4) and (1.2.5) at $r = a$ gives $c = 2\pi a^2 \sigma_0$, whence

$$V = \tfrac{2}{3}\pi\sigma_0(3a^2 - r^2); \qquad r \leqslant a.$$

Writing $r^2 = x^2 + y^2 + z^2$, it may be verified without difficulty that $\partial V/\partial x$, etc., are continuous at $r = a$.

1.3 Surface distribution of simple sources

A continuous distribution of simple sources extending over a Liapunov surface ∂B (not necessarily closed), and of surface density $\sigma(\mathbf{q})$ at $\mathbf{q} \in \partial B$, generates the simple-layer potential

$$V(\mathbf{p}) = \int_{\partial B} g(\mathbf{p}, \mathbf{q})\sigma(\mathbf{q})\,dq. \tag{1.3.1}$$

Roughly speaking a Liapunov surface (see Appendix 1) has a continuously varying tangent plane at each point, but it does not necessarily possess a curvature everywhere (Pogorzelski, 1966). Henceforth we shall assume, unless otherwise stated, that all surfaces are Liapunov surfaces. Such surfaces are slightly less general than Kellogg regular surfaces, and indeed Kellogg regularity suffices to ensure the fundamental existence-uniqueness theorems of harmonic function theory (Section 2.1). However, the restriction to Liapunov surfaces is necessary for formulations of boundary-value problems via potential theory (Section 2.2 et seq.). In particular, the potential (1.3.1) is continuous everywhere and has the asymptotic behaviour

$$V(\mathbf{p}) = |\mathbf{p}|^{-1} \int_{\partial B} \sigma(\mathbf{q})\,dq + O(|\mathbf{p}|^{-2}) \quad \text{as} \quad |\mathbf{p}| \to \infty. \tag{1.3.2}$$

It is differentiable to the second order and satisfies Laplace's equation, and it is therefore a harmonic function, everywhere except at ∂B. Provided that σ is Hölder continuous at $\mathbf{p} \in \partial B$, the tangential derivatives of V exist and are continuous at $\mathbf{p}$, but its normal derivatives are discontinuous (Smirnov, 1964). To examine these we erect a normal line to one side of ∂B at $\mathbf{p}$ and locate points on the normal by a variable n which increases moving away from ∂B. At any point on the normal other than the initial point, which belongs to ∂B,

$$\frac{\partial V(\mathbf{p})}{\partial n} = \int_{\partial B} \frac{\partial g(\mathbf{p}, \mathbf{q})}{\partial n} \sigma(\mathbf{q}) \, dq, \tag{1.3.3}$$

where $\partial g(\mathbf{p}, \mathbf{q})/\partial n$ denotes the derivative of g at $\mathbf{p}$ keeping $\mathbf{q}$ fixed. But at the initial point

$$\frac{\partial V(\mathbf{p})}{\partial n} = \int_{\partial B} \frac{\partial g(\mathbf{p}, \mathbf{q})}{\partial n} \sigma(\mathbf{q}) \, dq - 2\pi\sigma(\mathbf{p}); \qquad \mathbf{p} \in \partial B, \tag{1.3.4}$$

because of the singularity which arises in g when $\mathbf{q}$ coincides with the field point $\mathbf{p}$. An apparent indeterminacy occurs in $\partial g/\partial n$ when $\mathbf{q} = \mathbf{p}$, but its behaviour there may be uniquely obtained by a limiting process, e.g. as later in (1.4.19). It is often convenient to use the notation

$$\left. \begin{aligned} \frac{\partial V(\mathbf{p})}{\partial n} &= V'(\mathbf{p}) \\[2em] \frac{\partial g(\mathbf{p}, \mathbf{q})}{\partial n} &= g'(\mathbf{p}, \mathbf{q}) = g(\mathbf{q}, \mathbf{p})' \end{aligned} \right\} \tag{1.3.5}$$

for normal derivatives, in which case (1.3.4) appears as

$$V'(\mathbf{p}) = \int_{\partial B} g'(\mathbf{p}, \mathbf{q})\sigma(\mathbf{q}) \, dq - 2\pi\sigma(\mathbf{p}); \qquad \mathbf{p} \in \partial B. \tag{1.3.6}$$

There exist two distinct normals at $\mathbf{p}$, one on either side of ∂B. We shall adopt the convention that these two normals have equal status, in the sense that the relevant variables n_i, n_e both increase moving away from ∂B.

Correspondingly, we expand (1.3.6) into

$$V'_i(\mathbf{p}) = \int_{\partial B} g'_i(\mathbf{p}, \mathbf{q})\sigma(\mathbf{q})\, dq - 2\pi\sigma(\mathbf{p}); \qquad \mathbf{p} \in \partial B, \tag{1.3.7}$$

$$V'_e(\mathbf{p}) = \int_{\partial B} g'_e(\mathbf{p}, \mathbf{q})\sigma(\mathbf{q})\, dq - 2\pi\sigma(\mathbf{p}); \qquad \mathbf{p} \in \partial B. \tag{1.3.8}$$

Since $g(\mathbf{p}, \mathbf{q})$ remains continuous as $\mathbf{p}$ crosses ∂B, it follows that

$$g'_i(\mathbf{p}, \mathbf{q}) + g'_e(\mathbf{p}, \mathbf{q}) = 0; \qquad \mathbf{p}, \mathbf{q} \in \partial B, \tag{1.3.9}$$

whence

$$V'_i(\mathbf{p}) + V'_e(\mathbf{p}) = -4\pi\sigma(\mathbf{p}); \qquad \mathbf{p} \in \partial B. \tag{1.3.10}$$

This result is evidently a degenerate case of (1.1.8).

If ∂B is a closed surface, we must distinguish between the interior domain B_i enclosed by ∂B and the infinite domain B_e exterior to ∂B. Similarly we must distinguish between the interior simple-layer potential generated by σ and the exterior simple-layer potential generated by σ. As a simple exercise on this, suppose ∂B to be a closed spherical surface of radius a, with $\sigma = \sigma_0$ everywhere on ∂B, in which case V is radially symmetric everywhere. Hence, by virtue of (1.3.2),

$$V(r) = r^{-1}4\pi a^2\sigma_0; \qquad r \geqslant a,$$

$$V(r) = V(a) = 4\pi a\sigma_0; \qquad r \leqslant a.$$

It follows that

$$V'_e(a) = \left(\frac{dV}{dr}\right)_{r=a} = -4\pi\sigma_0,$$

$$V'_i(a) = \left(-\frac{dV}{dr}\right)_{r=a} = 0,$$

so that

$$V'_e(a) + V'_i(a) = -4\pi\sigma_0$$

in accordance with (1.3.10).

1.4 Surface distribution of double sources

A unit double source located at $\mathbf{q} \in \partial B$ generates the potential $g(\mathbf{p}, \mathbf{q})'_i$ at any other point $\mathbf{p}$, where $g(\mathbf{p}, \mathbf{q})'_i$ denotes the n_i-normal derivative of g at $\mathbf{q}$ keeping $\mathbf{p}$ fixed. This potential is a continuous function of $\mathbf{p}$, differentiable to all orders, and it satisfies Laplace's equation everywhere except at the source point $\mathbf{q}$. These properties characterise $g(\mathbf{p}, \mathbf{q})'_i$ as a harmonic function of $\mathbf{p}$ everywhere except at $\mathbf{p} = \mathbf{q}$. Formally $g(\mathbf{p}, \mathbf{q})'_i$ satisfies Poisson's equation

$$\nabla^2 g(\mathbf{p}, \mathbf{q})'_i = -4\pi\delta(\mathbf{p} - \mathbf{q})'_i \qquad (1.4.1)$$

everywhere, where the right-hand side of (1.4.1) may be given a consistent interpretation (Jones, 1966; Lighthill, 1958).

A discrete distribution of double sources, of strengths $\mu_1, \mu_2, \ldots, \mu_N$ located at $\mathbf{q}_1, \mathbf{q}_2, \ldots, \mathbf{q}_N$ on ∂B, generates the potential

$$W(\mathbf{p}) = \sum_{\alpha=1}^{N} g(\mathbf{p}, \mathbf{q}_\alpha)'_i \mu_\alpha \qquad (1.4.2)$$

at $\mathbf{p}$. This has essentially the same properties as $g(\mathbf{p}, \mathbf{q})'_i$, in particular it is a harmonic function everywhere except at $\mathbf{q}_1, \mathbf{q}_2, \ldots, \mathbf{q}_N$ and it formally satisfies Poisson's equation

$$\nabla^2 W(\mathbf{p}) = -4\pi \sum_{\alpha=1}^{N} \delta(\mathbf{p} - \mathbf{q}_\alpha)'_i \mu_\alpha \qquad (1.4.3)$$

everywhere.

A continuous distribution of double sources extending over ∂B (not necessarily closed), and of surface density $\mu(\mathbf{q})$ at $\mathbf{q} \in \partial B$, generates the double-layer potential

$$W(\mathbf{p}) = \int_{\partial B} g(\mathbf{p}, \mathbf{q})'_i \mu(\mathbf{q}) \, dq. \qquad (1.4.4)$$

This is continuous and differentiable to the second order, and satisfies Laplace's equation, and is therefore a harmonic function, everywhere except at ∂B, with asymptotic behaviour

$$W(\mathbf{p}) = O(|\mathbf{p}|^{-2}) \quad \text{as} \quad |\mathbf{p}| \to \infty. \tag{1.4.5}$$

To define its discontinuity properties at ∂B, let $\mathbf{p}_i$, $\mathbf{p}_e$ be points on the n_i, n_e normals respectively emanating from $\mathbf{p} \in \partial B$. Then

$$\lim_{\mathbf{p}_i \to \mathbf{p}} W(\mathbf{p}_i) = W(\mathbf{p}) + 2\pi\mu(\mathbf{p}), \tag{1.4.6}$$

$$\lim_{\mathbf{p}_e \to \mathbf{p}} W(\mathbf{p}_e) = W(\mathbf{p}) - 2\pi\mu(\mathbf{p}). \tag{1.4.7}$$

As regards the normal derivatives of W, the continuity of μ ensures that

$$\lim_{\mathbf{p}_i \to \mathbf{p}} \frac{\partial W(\mathbf{p}_i)}{\partial n_i} + \lim_{\mathbf{p}_e \to \mathbf{p}} \frac{\partial W(\mathbf{p}_e)}{\partial n_e} = 0; \qquad \mathbf{p} \in \partial B. \tag{1.4.8}$$

However, the separate limits necessarily exist only if $\mu \in C^{(2)}$ or, less restrictively, if μ is Hölder continuously differentiable at $\mathbf{p} \in \partial B$ (Günter, 1967). In this case the tangential derivatives also exist and satisfy the limiting relations

$$\left. \begin{aligned} \lim_{\mathbf{p}_i \to \mathbf{p}} \frac{\partial W(\mathbf{p}_i)}{\partial t_i} &= \frac{\partial W(\mathbf{p})}{\partial t} + 2\pi \frac{\partial \mu(\mathbf{p})}{\partial t} \\[2mm] \lim_{\mathbf{p}_e \to \mathbf{p}} \frac{\partial W(\mathbf{p}_e)}{\partial t_e} &= \frac{\partial W(\mathbf{p})}{\partial t} - 2\pi \frac{\partial \mu(\mathbf{p})}{\partial t} \end{aligned} \right\} \tag{1.4.9}$$

where $\partial/\partial t$, $\partial/\partial t_i$, $\partial/\partial t_e$ denote differentiation parallel to ∂B at $\mathbf{p}$, $\mathbf{p}_i$, $\mathbf{p}_e$ respectively.

If ∂B is a closed surface and $\mu = 1$,

$$W(\mathbf{p}) = \int_{\partial B} g(\mathbf{p}, \mathbf{q})'_i \, dq = 4\pi; \qquad \mathbf{p} \in B_i, \tag{1.4.10}$$

by virtue of the Gauss flux theorem (1.1.8). It then follows from (1.4.6) that

$$\int_{\partial B} g(\mathbf{p}, \mathbf{q})'_i \, dq = 2\pi; \qquad \mathbf{p} \in \partial B, \tag{1.4.11}$$

and from (1.4.7) that

$$\int_{\partial B} g(\mathbf{p}, \mathbf{q})'_i \, dq = 0; \qquad \mathbf{p} \in B_e. \tag{1.4.12}$$

Changing i into e in (1.4.4) yields the double-layer potential

$$W(\mathbf{p}) = \int_{\partial B} g(\mathbf{p}, \mathbf{q})'_e \mu(\mathbf{q}) \, dq, \tag{1.4.13}$$

which has the same value but the opposite sign to the potential (1.4.4) for every $\mathbf{p}$. Correspondingly, if ∂B is a closed surface and $\mu = 1$, the results (1.4.10)–(1.4.12) immediately yield

$$\int_{\partial B} g(\mathbf{p}, \mathbf{q})'_e \, dq = -4\pi; \qquad \mathbf{p} \in B_i, \tag{1.4.14}$$

$$\int_{\partial B} g(\mathbf{p}, \mathbf{q})'_e \, dq = 2\pi; \qquad \mathbf{p} \in \partial B, \tag{1.4.15}$$

$$\int_{\partial B} g(\mathbf{p}, \mathbf{q})'_e \, dq = 0; \qquad \mathbf{p} \in B_e. \tag{1.4.16}$$

Alternatively, we may enclose B_e by a large spherical surface ∂S and apply (1.1.8) to $\partial B + \partial S$. This gives

$$\int_{\partial B} g(\mathbf{p}, \mathbf{q})'_e \, dq + \int_{\partial S} g(\mathbf{p}, \mathbf{q})' \, dq = 4\pi; \qquad \mathbf{p} \in B_e,$$

where $g(\mathbf{p}, \mathbf{q})'$ denotes the inward normal derivative at $\mathbf{q} \in \partial S$, whence (1.4.16) follows since

$$\int_{\partial S} g(\mathbf{p}, \mathbf{q})' \, dq = 4\pi.$$

Similarly the result (1.4.15) follows from

$$\int_{\partial B} g(\mathbf{p}, \mathbf{q})'_e \, dq + \int_{\partial S} g(\mathbf{p}, \mathbf{q})' \, dq = 2\pi; \qquad \mathbf{p} \in \partial B,$$

since the second integral retains its previous value of 4π.

As an exercise on the preceding, suppose ∂B to be the closed spherical surface $x^2 + y^2 + z^2 - a^2 = 0$, with $\mathbf{p} = (0, 0, a)$; $\mathbf{q} = (r \sin\theta \cos\psi, r \sin\theta \sin\psi, r \cos\theta)_{r=a}$ where r, θ, ψ are the usual spherical polar co-ordinates. Now

$$g(\mathbf{p}, \mathbf{q}) = \{r^2 \sin^2\theta + (a - r\cos\theta)^2\}^{-\frac{1}{2}}_{r=a} = (r^2 + a^2 - 2ra\cos\theta)^{-\frac{1}{2}}_{r=a}$$

$$= \left(2a \sin\frac{\theta}{2}\right)^{-1}, \tag{1.4.17}$$

$$g(\mathbf{p}, \mathbf{q})'_e = \left(\frac{\partial g}{\partial r}\right)_{r=a} = \left(-4a^2 \sin\frac{\theta}{2}\right)^{-1}, \tag{1.4.18}$$

showing that

$$\lim_{\mathbf{q} \to \mathbf{p}} g(\mathbf{p}, \mathbf{q})'_e = \lim_{\theta \to 0} \left(-4a^2 \sin\frac{\theta}{2}\right)^{-1}. \tag{1.4.19}$$

Since $dq = a^2 \sin\theta \, d\theta \, d\psi$,

$$\int_{\partial B} g(\mathbf{p}, \mathbf{q})'_e \, dq = -\int_{\psi=0}^{\psi=2\pi} \int_{\theta=0}^{\theta=\pi} \left(4a^2 \sin\frac{\theta}{2}\right)^{-1} a^2 \sin\theta \, d\theta \, d\psi = -2\pi$$

in accordance with (1.4.15). It will be noted that

$$g(\mathbf{p}, \mathbf{q})'_e = -\frac{1}{2a} g(\mathbf{p}, \mathbf{q}), \tag{1.4.20}$$

which implies that $g(\mathbf{p}, \mathbf{q})$ and $g(\mathbf{p}, \mathbf{q})'_e$ have the same order of singularity for any surface which can be approximated locally by a sphere of contact. More generally, for a surface with principal radii of curvature at each point, arguments are available leading to the same conclusion (Duff, 1956). For a general Liapunov surface, not necessarily possessing a curvature at each point, it may be proved (Pogorzelski, 1966) that

$$|g(\mathbf{p}, \mathbf{q})'_e| \leqslant \frac{K}{|\mathbf{p} - \mathbf{q}|^{2-v}}; \qquad 0 \leqslant v \leqslant 1. \tag{1.4.21}$$

Clearly the weakest possible singularity occurs when $v = 1$, which covers the local spherical approximation (1.4.20) with $K = \max\left(\dfrac{1}{2a}\right)$ over ∂B.

1.5 Flux of normal derivatives

We are now in a position to integrate V'_i, V'_e over a closed surface ∂B. Formally

$$\int_{\partial B} V'_i(\mathbf{p})\, \mathrm{d}p = \int_{\partial B} \left\{ \int_{\partial B} g'_i(\mathbf{p}, \mathbf{q})\sigma(\mathbf{q})\, \mathrm{d}q \right\} \mathrm{d}p - 2\pi \int_{\partial D} \sigma(\mathbf{p})\, \mathrm{d}p, \qquad (1.5.1)$$

where

$$\int_{\partial B} \left\{ \int_{\partial B} g'_i(\mathbf{p}, \mathbf{q})\sigma(\mathbf{q})\, \mathrm{d}q \right\} \mathrm{d}p = \int_{\partial B} \sigma(\mathbf{q}) \left\{ \int_{\partial B} g'_i(\mathbf{p}, \mathbf{q})\, \mathrm{d}p \right\} \mathrm{d}q$$

$$= 2\pi \int_{\partial B} \sigma(\mathbf{q})\, \mathrm{d}q \qquad (1.5.2)$$

on interchanging $\mathbf{p}, \mathbf{q}$ in (1.4.11). The order of the double integration has been interchanged by applying Fubini's theorem (Petrovsky, 1971) since

$$\int_{\partial B} \int_{\partial B} |g'_i(\mathbf{p}, \mathbf{q})\sigma(\mathbf{q})|\, \mathrm{d}p\, \mathrm{d}q$$

exists for a Liapunov surface, as follows by a fairly straightforward application of (1.4.21). Hence the right-hand side of (1.5.1) equals zero, showing that

$$\int_{\partial B} V'_i(\mathbf{p})\, \mathrm{d}p = 0 \qquad (1.5.3)$$

in accordance with (1.1.9).

Similarly, changing i into e in (1.5.1) gives

$$\int_{\partial B} V'_e(\mathbf{p})\, \mathrm{d}p = -4\pi \int_{\partial B} \sigma(\mathbf{p})\, \mathrm{d}p \qquad (1.5.4)$$

in accordance with (1.1.8). Alternatively, we may enclose ∂B by a large spherical surface ∂S and apply (1.1.9) to $\partial B + \partial S$. This gives

$$\int_{\partial B} V'_e(\mathbf{p}) \, dp + \int_{\partial S} V'(\mathbf{p}) \, dp = 0 \qquad (1.5.5)$$

where V' denotes the inward normal derivative at $\mathbf{p} \in \partial S$, whence (1.5.4) follows since

$$\int_{\partial S} V'(\mathbf{p}) \, dp = 4\pi \int_{\partial B} \sigma(\mathbf{p}) \, dp,$$

by an application of (1.3.2).

Our analysis may be extended to a finite domain B bounded internally by a closed surface ∂B_1 and externally by a closed surface ∂B_0. We introduce continuous source distributions over ∂B_1 and ∂B_0 which generate the simple-layer potential

$$V(\mathbf{p}) = \int_{\partial B_1} g(\mathbf{p}, \mathbf{q}) \sigma(\mathbf{q}) \, dq + \int_{\partial B_0} g(\mathbf{p}, \mathbf{q}) \sigma(\mathbf{q}) \, dq \qquad (1.5.6)$$

at any point $\mathbf{p}$ of space. This potential exists and is continuous everywhere, and it defines distinct harmonic functions within B, within the domain enclosed by ∂B_1, and within the infinite domain exterior to ∂B_0. Clearly

$$V(\mathbf{p}) = \left\{ \int_{\partial B_1} \sigma(\mathbf{q}) \, dq + \int_{\partial B_0} \sigma(\mathbf{q}) \, dq \right\} |\mathbf{p}|^{-1} + O(|\mathbf{p}|^{-2}) \quad \text{as} \quad |\mathbf{p}| \to \infty. \ (1.5.7)$$

The normal derivative of V directed into B at $\mathbf{p} \in \partial B_1$ is

$$V'(\mathbf{p}) = \frac{\partial}{\partial n_e} \int_{\partial B_1} g(\mathbf{p}, \mathbf{q}) \sigma(\mathbf{q}) \, dq + \frac{\partial}{\partial n_e} \int_{\partial B_0} g(\mathbf{p}, \mathbf{q}) \sigma(\mathbf{q}) \, dq; \qquad \mathbf{p} \in \partial B_1$$

$$= \left\{ \int_{\partial B_1} g'_e(\mathbf{p}, \mathbf{q}) \sigma(\mathbf{q}) \, dq - 2\pi\sigma(\mathbf{p}) \right\} + \int_{\partial B_0} g'_e(\mathbf{p}, \mathbf{q}) \sigma(\mathbf{q}) \, dq;$$

$$\mathbf{p} \in \partial B_1, \qquad (1.5.8)$$

and therefore

$$\int_{\partial B_1} V'(\mathbf{p}) \, dp = -4\pi \int_{\partial B_1} \sigma(\mathbf{p}) \, dp, \qquad (1.5.9)$$

bearing in mind

$$\int_{\partial B_1} \left\{ \int_{\partial B_1} g'_e(\mathbf{p}, \mathbf{q})\sigma(\mathbf{q})\, dq - 2\pi\sigma(\mathbf{p}) \right\} dp = -4\pi \int_{\partial B_1} \sigma(\mathbf{p})\, dp,$$

$$\int_{\partial B_1} \left\{ \int_{\partial B_0} g'_e(\mathbf{p}, \mathbf{q})\sigma(\mathbf{q})\, dq \right\} dp = \int_{\partial B_0} \sigma(\mathbf{q}) \left\{ \int_{\partial B_1} g'_e(\mathbf{p}, \mathbf{q})\, dp \right\} dq = 0.$$

The normal derivative of V directed into B at $\mathbf{p} \in \partial B_0$ is

$$V'(\mathbf{p}) = \frac{\partial}{\partial n_i} \int_{\partial B_1} g(\mathbf{p}, \mathbf{q})\sigma(\mathbf{q})\, dq + \frac{\partial}{\partial n_i} \int_{\partial B_0} g(\mathbf{p}, \mathbf{q})\sigma(\mathbf{q})\, dq; \qquad \mathbf{p} \in \partial B_0$$

$$= \int_{\partial B_1} g'_i(\mathbf{p}, \mathbf{q})\sigma(\mathbf{q})\, dq + \left\{ \int_{\partial B_0} g'_i(\mathbf{p}, \mathbf{q})\sigma(\mathbf{q})\, dq - 2\pi\sigma(\mathbf{p}) \right\};$$

$$\mathbf{p} \in \partial B_0, \qquad (1.5.10)$$

and therefore

$$\int_{\partial B_0} V'(\mathbf{p})\, dp = 4\pi \int_{\partial B_1} \sigma(\mathbf{p})\, dp, \qquad (1.5.11)$$

bearing in mind

$$\int_{\partial B_0} \left\{ \int_{\partial B_1} g'_i(\mathbf{p}, \mathbf{q})\sigma(\mathbf{q})\, dq \right\} dp = \int_{\partial B_1} \sigma(\mathbf{q}) \left\{ \int_{\partial B_0} g'_i(\mathbf{p}, \mathbf{q})\, dp \right\} dq = 4\pi \int_{\partial B_1} \sigma(\mathbf{q})\, dq,$$

$$\int_{\partial B_0} \left\{ \int_{\partial B_0} g'_i(\mathbf{p}, \mathbf{q})\sigma(\mathbf{q})\, dq - 2\pi\sigma(\mathbf{p}) \right\} dp = 0.$$

Accordingly

$$\int_{\partial B_1} V'(\mathbf{p})\, dp + \int_{\partial B_0} V'(\mathbf{p})\, dp = 0$$

as expected. There is no difficulty in proving that

$$\int_{\partial B_0} V'(\mathbf{p})\, dp = -4\pi \left\{ \int_{\partial B_1} \sigma(\mathbf{p})\, dp + \int_{\partial B_0} \sigma(\mathbf{p})\, dp \right\},$$

$$\int_{\partial B_1} V'(\mathbf{p})\, dp = 0.$$

when $V'(\mathbf{p})$ denotes the normal derivative of V directed away from B at $\mathbf{p} \in \partial B_0, \partial B_1$ respectively.

1.6 Vortices

An important role in fluid mechanics is played by the open double-layer sheet of uniform source density. Here W jumps by the same amount $4\pi\mu$ on crossing ∂B at any point. However, the physically significant quantity ∇W, i.e. the fluid velocity $\mathbf{v}$, remains continuous since $\partial W/\partial n$, $\partial W/\partial t$ remain continuous if μ is constant. Closely associated with $\mathbf{v}$ is the vorticity vector $\boldsymbol{\omega} = \nabla \wedge \mathbf{v}$, and clearly $\boldsymbol{\omega} = 0$ everywhere except, possibly, on the boundary contour L of ∂B. Within the context of fluid mechanics, L is a vortex line but may preferably be regarded as the limit of a closed vortex tube L^* of finite cross-section. To gain an insight into $\boldsymbol{\omega}$ in L^*, we introduce the circulation integral

$$\oint \mathbf{v} \cdot d\mathbf{s} = \oint \nabla W \cdot d\mathbf{s} = 4\pi\mu \tag{1.6.1}$$

where $\bigcirc$ denotes any irreducible circuit around L^* passing once through ∂B. Assuming that the cross-section of L^* is sufficiently small for $\boldsymbol{\omega}$ to remain constant across it, we obtain

$$\boldsymbol{\omega} \cdot d\mathbf{S} = \oint \mathbf{v} \cdot d\mathbf{s} = 4\pi\mu \tag{1.6.2}$$

by an application of Stokes' theorem, where $d\mathbf{S}$ denotes the elementary area vector at the appropriate section of L^*. Clearly $\boldsymbol{\omega} \cdot d\mathbf{S}$ has a constant value along L^* which defines the vortex strength. According to the Biot–Savart

formula, (Thwaites, 1960)

$$\mathbf{v} = \nabla \wedge \frac{1}{4\pi} \int_{L*} \frac{\omega(\mathbf{q}) \, dQ}{|\mathbf{p} - \mathbf{q}|} = \frac{\omega \cdot d\mathbf{S}}{4\pi} \int_{L*} \frac{d\mathbf{q} \wedge (\mathbf{p} - \mathbf{q})}{|\mathbf{p} - \mathbf{q}|^3} \qquad (1.6.3)$$

where dQ denotes the volume element and $d\mathbf{q}$ the vector length of arc at $\mathbf{q} \in L^*$. This formula provides a preferred alternative to ∇W in problems where L^* rather than ∂B appears as the main focus of interest.

In electromagnetic theory we envisage L^* as a conducting filament through which there passes a current of strength $\omega \cdot d\mathbf{S}$. Also we interpret ω as the current vector and $\mathbf{v}$ as the magnetic field generated by the current. Clearly its effect is equivalent to that of any magnetic sheet of uniform moment $(1/4\pi)\omega \cdot d\mathbf{S}$ contained by the filament.

2

Representation of Harmonic Functions by Potentials

2.1 Fundamental existence-uniqueness theorems

A function ϕ is said to be harmonic within a three-dimensional domain B_i, bounded by a closed surface ∂B, if it satisfies the following conditions:
 (a) ϕ is continuous in $B_i + \partial B$,
 (b) ϕ is differentiable to at least the second order in B_i,
 (c) ϕ satisfies Laplace's equation in B_i.
 If ∂B is a Liapunov surface,[†] it is possible to determine ϕ throughout B_i in terms of suitably prescribed information over ∂B. For example, given arbitrary continuous values of ϕ over ∂B, there exists a unique ϕ in B_i which assumes these values. This is the Dirichlet existence theorem of harmonic function theory. An immediate consequence of this theorem is that $\phi = 1$ over ∂B implies $\phi = 1$ in $B_i + \partial B$. Also, given arbitrary continuous values of $\phi_i'(= \partial\phi/\partial n_i)$ over ∂B, which satisfy the Gauss condition

$$\int_{\partial B} \phi_i'(\mathbf{q})\, dq = 0,$$

/* i.e. no nett source of 'potential' ϕ */

there exists a unique (up to an arbitrary additive constant) ϕ in B_i whose normal derivative assumes these values. This is the Neumann existence theorem of harmonic function theory. An immediate consequence of this theorem is that $\phi_i' = 0$ over ∂B implies $\phi = C$ (an arbitrary constant) in $B_i + \partial B$.

Harmonic functions may also exist within the infinite domain B_e exterior to ∂B. The preceding fundamental theorems remain valid provided that

18

† A smooth surface defined in Appendix 1

$\phi = O(r^{-1})$ as $r \to \infty$ (i.e. ϕ is regular in B_e). Thus, given arbitrary continuous values of ϕ over ∂B, there exists a unique regular ϕ in B_e which assumes these values (exterior Dirichlet existence theorem). An immediate consequence of this theorem is that $\phi = 0$ over ∂B implies the (regular) harmonic function $\phi = 0$ in $B_e + \partial B$. Also, given arbitrary continuous values of ϕ'_e over ∂B, there exists a unique regular ϕ in $B_e + \partial B$ whose normal derivative assumes these values (exterior Neumann existence theorem). An immediate consequence of this theorem is that $\phi'_e = 0$ over ∂B implies the (regular) harmonic function $\phi = 0$ in $B_e + \partial B$. The Gauss condition $\int_{\partial B} \phi'_e(\mathbf{q}) \, dq = 0$ need not be satisfied by ϕ'_e since $\int_{\partial B} \phi'_e(\mathbf{q}) \, dq$ is balanced by a compensating flux at infinity. More precisely, following (1.5.5), we have

$$\int_{\partial B} \phi'_e(\mathbf{q}) \, dq + \int_{\partial S} \phi'(\mathbf{q}) \, dq = 0 \tag{2.1.1}$$

where

$$\int_{\partial S} \phi'(\mathbf{q}) \, dq = O(1) \quad \text{if} \quad \phi = O(r^{-1}) \quad \text{as} \quad r \to \infty. \tag{2.1.2}$$

In the event that $\phi = O(r^{-2})$ as $r \to \infty$ the flux over ∂S vanishes† and we deduce $\int_{\partial B} \phi'_e(\mathbf{q}) \, dq = 0$. Conversely, if this latter flux happens to vanish, it follows that $\phi = O(r^{-2})$ as $r \to \infty$.

The Dirichlet and Neumann boundary conditions are particular cases of a prescribed linear relation

$$\boxed{\alpha \phi + \beta \frac{\partial \phi}{\partial n} = \gamma} \tag{2.1.3}$$

between ϕ and $\partial \phi / \partial n$ at each point of ∂B. Denoting by $C(\partial B)$ the class of all continuous functions over ∂B, the Dirichlet condition is defined by —

$$\alpha = 1, \quad \beta = 0, \quad \gamma \in C(\partial B), \tag{2.1.4}$$

and the Neumann condition is defined by —

$$\alpha = 0, \quad \beta = 1, \quad \gamma \in C(\partial B). \tag{2.1.5}$$

† An integral which enters into an equation where all the other terms are fixed must necessarily have a constant value. If the behaviour is $O(r^{-1})$ as $r \to \infty$, the constant must be zero.

B

An existence–uniqueness theorem is also available for the Robin boundary condition of heat conduction (Kellogg, 1929) which is defined by

$$\alpha < 0, \qquad \beta = 1, \qquad \alpha \in H(\partial B), \qquad \gamma \in H(\partial B) \qquad (2.1.6)$$

where $H(\partial B)$ denotes the class of all Hölder continuous functions[+] over ∂B. The inequality $\alpha < 0$ has been chosen to be consistent with our convention that $\partial \phi / \partial n$ signifies the normal derivative directed into the domain B under consideration. Finally, an existence-uniqueness theorem is available for the difficult mixed boundary conditions defined by ——

$$
\left.
\begin{aligned}
\alpha = 1, \qquad \beta = 0, \qquad &\gamma \in C(\partial B_1), \\[2mm]
\alpha = 0, \qquad \beta = 1, \qquad &\gamma \in C(\partial B_2), \\[2mm]
\partial B_1 + \partial B_2 = \partial B, \qquad &
\end{aligned}
\right\} \qquad (2.1.7)
$$

where γ may jump on passing from ∂B_1 to ∂B_2 (Tsuji, 1959).

2.2 Boundary-value problems

Each existence theorem raises a corresponding boundary-value problem, i.e. that of constructing the appropriate harmonic function throughout the domain in question. An effective method of attack is to represent ϕ by a simple-layer or a double-layer potential generated by continuous source-distributions over ∂B, henceforth restricted to be a Liapunov surface. Since these potentials are harmonic functions in B, they will be identical with ϕ in B (though not necessarily in $B + \partial B$) provided they satisfy the boundary conditions prescribed for ϕ. This procedure leads to the formulation of integral equations which define the source densities concerned. The boundary-value problems accordingly reduce to formal mathematical problems, viz. those of ascertaining that the integral equations possess solutions and of computing them. Generally speaking the achievement of exact or even approximately exact analytical solutions is out of the question. Further progress can only be made by discretising the equations and computing numerical solutions which generate the potentials to a tolerable accuracy. Numerical potential theory, with applications to various interior and exterior problems, is treated in Part II.

[+] A stronger than normal form of continuity - Appendix 2

A conceptual disadvantage of simple-layer and double-layer potentials is the introduction of formal source densities which usually bear no relation to the problem. As we shall see in Chapter 3, this defect is removed in Green's formula, where ϕ, $\partial\phi/\partial n$ over ∂B play the role of source densities which generate ϕ throughout B. Nevertheless the representation of ϕ by a single potential is sufficiently important both in theory and in practice to warrant an independent treatment.

In the sequel we shall be concerned with integral operators generated by the kernels

$$k(\mathbf{p}, \mathbf{q}) = g(\mathbf{p}, \mathbf{q}), \qquad g(\mathbf{p}, \mathbf{q})', \qquad g'(\mathbf{p}, \mathbf{q}) \tag{2.2.1}$$

where $\mathbf{p}, \mathbf{q} \in \partial B$. These have the weak singularity

$$k(\mathbf{p}, \mathbf{q}) = O\left(\frac{1}{|\mathbf{p} - \mathbf{q}|}\right) \qquad \text{as } |\mathbf{p} - \mathbf{q}| \to 0, \tag{2.2.2}$$

as implied by (1.4.21) with $v = 1$. Strictly speaking such kernels fall outside the scope of the classical Fredholm kernels, which must be square integrable functions over ∂B (Smithies, 1958; Shilov, 1965). Nevertheless the classical theorems which hold for Fredholm integral equations of the second kind remain valid for weakly singular kernels (Pogorzelski, 1966; Mikhlin, 1970). It may be remarked that the modern theory of Fredholm integral equations appears as the theory of completely continuous operators in the Hilbert space $L_2(\partial B)$. The integral operators generated by the kernels (2.2.1) share this property. As regards Fredholm integral equations of the first kind, little general theory is available. (The underlying reason is indicated by Courant and Hilbert (1953).) Fortunately, within the present context, integral equations of the first kind may usually be transformed into integral equations of the second kind, so rendering them amenable to an available theory.

2.3 Electrostatic capacitance

A useful introduction to boundary-value problems is provided by the capacitance problem of electrostatics. In its simplest form we picture ∂B as a closed, perfectly conducting surface to which electric charges are introduced. These may be regarded mathematically as a simple-layer distribution over

∂B, of Hölder continuous density $\lambda(\mathbf{q})$ at $\mathbf{q} \in \partial B$, which generates the potential

$$V(\mathbf{p}) = \int_{\partial B} g(\mathbf{p}, \mathbf{q})\, \lambda(\mathbf{q})\, dq \qquad (2.3.1)$$

everywhere. In equilibrium V remains constant over ∂B, i.e. $V = 1$ without loss of generality, so yielding the boundary relation

$$\int_{\partial B} g(\mathbf{p}, \mathbf{q})\, \lambda(\mathbf{q})\, dq = 1; \qquad \mathbf{p} \in \partial B \qquad (2.3.2)$$

which must be satisfied by λ. This is an integral equation of the first kind for λ, corresponding to a Dirichlet formulation of the capacitance problem. We shall prove that a unique solution of (2.3.2) exists, characterised by $\lambda > 0$, thereby allowing us to introduce the essentially positive quantity

$$\kappa = \int_{\partial B} \lambda(\mathbf{q})\, dq > 0 \qquad (2.3.3)$$

associated with ∂B. This quantity κ measures the electrostatic capacitance of ∂B.

Equation (2.3.2) by no means exhausts all the properties of λ. Since $V = 1$ over ∂B, $V = 1$ in $B_i + \partial B$ by the interior Dirichlet existence theorem. Hence $V_i' = 0$ over ∂B, so yielding the boundary relation

$$\boxed{\int_{\partial B} g_i'(\mathbf{p}, \mathbf{q})\, \lambda(\mathbf{q})\, dq - 2\pi\lambda(\mathbf{p}) = 0; \qquad \mathbf{p} \in \partial B.} \qquad (2.3.4)$$

This is a homogeneous integral equation of the second kind for λ, corresponding to a Neumann formulation of the capacitance problem. (An alternative derivation of (2.3.4) appears in Appendix 3.) According to classical Fredholm theory, this equation has a non-trivial solution if its adjoint equation

$$\int_{\partial B} g(\mathbf{p}, \mathbf{q})_i'\, \mu(\mathbf{q})\, dq - 2\pi\mu(\mathbf{p}) = 0; \qquad \mathbf{p} \in \partial B \qquad (2.3.5)$$

has a non-trivial solution. Now $\mu = 1$ satisfies (2.3.5) by virtue of (1.4.11), so demonstrating the existence of a corresponding non-trivial solution of (2.3.4).

If Λ is such a solution, it generates the interior potential $\int_{\partial B} g(\mathbf{p}, \mathbf{q}) \Lambda(\mathbf{q}) \, dq$ characterised by

$$\frac{\partial}{\partial n_i} \int_{\partial B} g(\mathbf{p}, \mathbf{q}) \Lambda(\mathbf{q}) \, dq = 0; \qquad \mathbf{p} \in \partial B, \tag{2.3.6}$$

whence, from the interior Neumann existence theorem,

$$\int_{\partial B} g(\mathbf{p}, \mathbf{q}) \Lambda(\mathbf{q}) \, dq = c \text{ (a constant)}; \qquad \mathbf{p} \in \partial B. \tag{2.3.7}$$

Provided that $c \neq 0$, it follows from (2.3.7) that $\lambda = \Lambda/c$ satisfies (2.3.2).

If, hypothetically $c = 0$, then Λ generates an exterior potential characterised by

(a)
$$\int_{\partial B} g(\mathbf{p}, \mathbf{q}) \Lambda(\mathbf{q}) \, dq = 0; \qquad \mathbf{p} \in \partial B, \tag{2.3.8}$$

(b)
$$\int_{\partial B} g(\mathbf{p}, \mathbf{q}) \Lambda(\mathbf{q}) \, dq = O(|\mathbf{p}|^{-1}) \qquad \text{as} \qquad |\mathbf{p}| \to \infty,$$

so implying

$$\int_{\partial B} g(\mathbf{p}, \mathbf{q}) \Lambda(\mathbf{q}) \, dq = 0; \qquad \mathbf{p} \in B_e + \partial B,$$

by the exterior Dirichlet existence theorem. Also, of course,

$$\int_{\partial B} g(\mathbf{p}, \mathbf{q}) \Lambda(\mathbf{q}) \, dq = 0; \qquad \mathbf{p} \in B_i + \partial B.$$

Therefore, by virtue of (1.3.10), $\Lambda = 0$. Five deductions now follow in order. First $c \neq 0$ since $\Lambda \neq 0$. Secondly, equation (2.3.8) only admits the trivial solution $\Lambda = 0$. Thirdly, equation (2.3.2) has a unique solution. Fourthly, equation (2.3.4) has only one independent non-trivial solution. (An alternative proof of this is provided by Duff (1956).) Fifthly, equation (2.3.5) has only one independent non-trivial solution.

By virtue of (2.3.2) the closed surface ∂B is an equipotential of V. Hence

V'_e has the same sign over ∂B. Also $V'_i = 0$ over ∂B. Therefore, since

$$\lambda = -\frac{1}{4\pi}(V'_e + V'_i) = -\frac{1}{4\pi} V'_e, \qquad (2.3.9)$$

λ has the same (positive) sign over ∂B. (A similar line of argument can be used to prove that a harmonic function does not attain a maximum or a minimum at any interior point of $B + \partial B$.) For example, taking ∂B to be a sphere of radius a, equation (2.3.2) by symmetry must admit the solution $\lambda(\mathbf{q}) = \lambda_0$ (a constant) and it may therefore be written

$$\lambda_0 \int_{\partial B} g(\mathbf{p}, \mathbf{q})\, dq = 1; \mathbf{p} \in \partial B. \qquad (2.3.10)$$

Now

$$\int_{\partial B} g(\mathbf{p}, \mathbf{q})\, dq = 4\pi a; \mathbf{p} \in \partial B \qquad (2.3.11)$$

for a sphere of radius a, whence $\lambda_0 = 1/4\pi a > 0$. This evidently generates the potentials

$$V(r) = r^{-1} 4\pi a^2 \frac{1}{4\pi a} = \frac{a}{r}; \qquad r \geqslant a,$$

$$V(r) = V(a) = 1; \qquad r \leqslant a,$$

which are seen to be equal over ∂B.

2.4 Neumann problems

Given continuous values of ϕ'_i over ∂B, there exists an appropriate ϕ in $B_i + \partial B$ for which we try the representation

$$\phi(\mathbf{p}) = \int_{\partial B} g(\mathbf{p}, \mathbf{q})\, \sigma(\mathbf{q})\, dq; \qquad \mathbf{p} \in B_i + \partial B, \qquad (2.4.1)$$

where σ is a source density to be determined. This yields the boundary relation

$$\phi_i'(\mathbf{p}) = \int_{\partial B} g_i'(\mathbf{p}, \mathbf{q})\, \sigma(\mathbf{q})\, \mathrm{d}q - 2\pi\sigma(\mathbf{p}); \qquad \mathbf{p} \in \partial B, \qquad (2.4.2)$$

which constitutes a Fredholm integral equation of the second kind for σ in terms of ϕ_i' The special case $\phi_i' = 0$, corresponding with the homogeneous equation (2.3.4), has already been examined. According to Fredholm theory, there exists a solution of (2.4.2) if, and only if,

$$\int_{\partial B} \mu(\mathbf{p})\, \phi_i'(\mathbf{p})\, \mathrm{d}p = 0 \qquad (2.4.3)$$

where μ satisfies the adjoint homogeneous equation (2.3.5). Since $\mu = 1$, condition (2.4.3) becomes the expected Gauss condition $\int_{\partial B} \phi_i'(\mathbf{p})\, \mathrm{d}p = 0$. Once a particular solution σ_0 of (2.4.2) has been achieved, the general solution appears as

$$\sigma_0 + k\lambda \qquad (2.4.4)$$

where k is an arbitrary constant and λ satisfies (2.3.4) subject to the normalisation (2.3.2). We utilise (2.4.4) to generate the class of potentials

$$\int_{\partial B} g(\mathbf{p}, \mathbf{q})\, \sigma_0(\mathbf{q})\, \mathrm{d}q + k \int_{\partial B} g(\mathbf{p}, \mathbf{q})\, \lambda(\mathbf{q})\, \mathrm{d}q; \qquad \mathbf{p} \in B_i + \partial B,$$

i.e.

$$\int_{\partial B} g(\mathbf{p}, \mathbf{q})\, \sigma_0(\mathbf{q})\, \mathrm{d}q + k; \qquad \mathbf{p} \in B_i + \partial B, \qquad (2.4.5)$$

each characterised by the same normal derivative ϕ_i' over ∂B.

Given continuous values of ϕ_e' over ∂B, there exists a unique regular ϕ in $B_e + \partial B$ whose normal derivative assumes these values. We try the representation

$$\phi(\mathbf{p}) = \int_{\partial B} g(\mathbf{p}, \mathbf{q})\, \sigma(\mathbf{q})\, \mathrm{d}q; \qquad \mathbf{p} \in B_e + \partial B, \qquad (2.4.6)$$

which provides acceptable behaviour at infinity and yields the boundary relation

$$\phi'_e(\mathbf{p}) = \int_{\partial B} g'_e(\mathbf{p}, \mathbf{q})\, \sigma(\mathbf{q})\, dq - 2\pi\sigma(\mathbf{p}); \qquad \mathbf{p} \in \partial B. \qquad (2.4.7)$$

This is a Fredholm integral equation of the second kind for σ in terms of ϕ'_e, with a homogeneous component

$$\int_{\partial B} g'_e(\mathbf{p}, \mathbf{q})\, \sigma(\mathbf{q})\, dq - 2\pi\sigma(\mathbf{p}) = 0; \qquad \mathbf{p} \in \partial B, \qquad (2.4.8)$$

and accompanying adjoint

$$\int_{\partial B} g(\mathbf{p}, \mathbf{q})'_e\, \tau(\mathbf{q})\, dq - 2\pi\tau(\mathbf{p}) = 0; \qquad \mathbf{p} \in \partial B. \qquad (2.4.9)$$

Unlike the interior homogeneous equation (2.3.4), equation (2.4.8) has no non-trivial solution. For if Λ satisfies (2.4.8), it generates an exterior potential characterised by

(a) $\qquad \dfrac{\partial}{\partial n_e} \displaystyle\int_{\partial B} g(\mathbf{p}, \mathbf{q})\, \Lambda(\mathbf{q})\, dq = 0; \qquad \mathbf{p} \in \partial B,$

(b) $\qquad \displaystyle\int_{\partial B} g(\mathbf{p}, \mathbf{q})\, \Lambda(\mathbf{q})\, dq = O(|\mathbf{p}|^{-1}) \qquad$ as $\qquad |\mathbf{p}| \to \infty.$

Therefore, by the exterior Neumann existence theorem, it follows that

$$\int_{\partial B} g(\mathbf{p}, \mathbf{q})\, \Lambda(\mathbf{q})\, dq = 0; \qquad \mathbf{p} \in B_e + \partial B. \qquad (2.4.10)$$

Also, by continuation from (2.4.10), we have

$$\int g(\mathbf{p}, \mathbf{q})\, \Lambda(\mathbf{q})\, dq = 0; \qquad \mathbf{p} \in B_i + \partial B,$$

whence, from (1.3.10), $\Lambda = 0$. An immediate deduction is that equation (2.4.9) has only the solution $\tau = 0$. Replacing μ by τ and ϕ_i' by ϕ_e' in (2.4.3), shows that equation (2.4.7) has a unique solution for arbitrary continuous values of ϕ_e'. A useful check on this solution σ is provided by the relation (1.5.4), i.e.

$$\int_{\partial B} \phi_e'(\mathbf{p}) \, dp = -4\pi \int_{\partial B} \sigma(\mathbf{p}) \, dp. \tag{2.4.11}$$

With σ known, we may generate ϕ anywhere in B_e and on ∂B itself utilising (2.4.6).

Equation (2.4.7) has an immediate application to the theory of potential fluid motion. Here we regard ∂B as a fixed, rigid obstacle which perturbs a steady flow defined by some given velocity potential ψ. If ϕ is the perturbation potential, then we require $\phi = O(r^{-1})$ at most as $r \to \infty$, since the perturbation decays at infinity. This behaviour is ensured by the representation (2.4.6). Also the total velocity potential is $\phi + \psi$, so that the total velocity is $\nabla(\phi + \psi)$. This has a zero component normal to ∂B, i.e.

$$\nabla(\phi + \psi) \cdot \mathbf{n}_e = \phi_e' + \psi_e' = 0 \quad \text{over} \quad \partial B$$

where $\mathbf{n}_e$ denotes the exterior unit normal vector at $\mathbf{p} \in \partial B$. Hence

$$\phi_e' = -\psi_e' \quad \text{over} \quad \partial B, \tag{2.4.12}$$

so providing the left-hand side of equation (2.4.7). It may be observed that

$$\int_{\partial B} \phi_e'(\mathbf{p}) \, dp = -\int_{\partial B} \psi_e'(\mathbf{p}) \, dp = \int_{\partial B} \psi_i'(\mathbf{p}) \, dp = 0, \tag{2.4.13}$$

bearing in mind (2.4.12) and the fact that ψ is continuous everywhere. Accordingly, from (2.4.11) and (1.3.2), or on more general grounds from (2.1.2) et seq.,

$$\phi = O(r^{-2}) \quad \text{as} \quad r \to \infty. \tag{2.4.14}$$

The perturbation of a uniform electrostatic field by a dielectric body may also be formulated as a Neumann boundary-value problem. We regard B_i as a medium of dielectric constant K_i, embedded in an infinite medium B_e of dielectric constant K_e, which transmits a uniform electrostatic field $\nabla\psi$. If ϕ_i, ϕ_e are the perturbations of ψ in B_i, B_e respectively, they must satisfy three distinct conditions (Jeans, 1920). First $\phi_e = O(r^{-1})$ at most as $r \to \infty$ for the same reason as previously, which behaviour is ensured by the representation (2.4.6). Secondly, the total potential must be continuous across ∂B,

i.e.
$$\psi + \phi_i = \psi + \phi_e \quad \text{over} \quad \partial B,$$

i.e.
$$\phi_i = \phi_e \quad \text{over} \quad \partial B, \tag{2.4.15}$$

which is ensured by introducing the representation (2.4.1) for ϕ_i with the same σ as in (2.4.6). Thirdly, the electric displacement vector must be continuous across ∂B,

i.e.
$$K_i(\psi_i' + \phi_i') + K_e(\psi_e' + \phi_e') = 0 \quad \text{over} \quad \partial B,$$

i.e.
$$K_i\phi_i' + K_e\phi_e' + (K_i - K_e)\psi_i' = 0 \quad \text{over} \quad \partial B, \tag{2.4.16}$$

yielding the boundary relation

$$(K_i - K_e)\int_{\partial B} g_i'(\mathbf{p}, \mathbf{q})\,\sigma(\mathbf{q})\,dq - 2\pi(K_i + K_e)\,\sigma(\mathbf{p}) = (K_e - K_i)\psi_i'(\mathbf{p});$$

$$\mathbf{p} \in \partial B, \tag{2.4.17}$$

on remembering that $g_i' + g_e' = 0$. This is a Fredholm integral equation of the second kind for σ in terms of ψ_i', which has a unique solution (see Appendix 3) except for the case $K_i = \infty$ (formally equivalent to $K_e = 0$) corresponding to an electrostatic conductor *in vacuo*.

Integrating both sides of (2.4.16), we find without difficulty that

$$K_e \int_{\partial B} \phi_e'(\mathbf{p})\,dp = 0 \tag{2.4.18}$$

in line with (2.4.13). Alternatively we may integrate both sides of (2.4.17) with respect to $\mathbf{p}$ to obtain the equivalent result

$$-4\pi K_e \int_{\partial B} \sigma(\mathbf{q})\,dq = 0. \tag{2.4.19}$$

The argument breaks down when $K_e = 0$, in which case equation (2.4.17)

reduces to the form (2.4.2) with ϕ_i' replaced by $-\psi_i'$. If so we recover the class
of solutions (2.4.4) including that defined by

$$\int_{\partial B} \sigma_0(\mathbf{p})\,dp + k \int_{\partial B} \lambda(\mathbf{p})\,dp = 0. \tag{2.4.20}$$

This particular solution generates a ϕ_e characterised by the asymptotic
behaviour (2.4.14) at infinity. Also, by reference to (2.4.16), it generates a ϕ_i
with the property

$$\phi_i + \psi = c\,(\text{a constant}) \quad \text{in} \quad B_i \mid \partial B, \tag{2.4.21}$$

i.e.

$$\phi_i + (\psi - c) = 0 \quad \text{in} \quad B_i + \partial B \tag{2.4.22}$$

for an earthed conductor since we are always at liberty to replace ψ by $\psi - c$.
Relation (2.4.22) applied to ∂B provides a Dirichlet formulation of the problem, in which c is initially unknown and is balanced by the side condition
$\int_{\partial B} \sigma(\mathbf{q})\,dq = 0$.

2.5 Dirichlet problems: simple-layer formulation

Given continuous values of ϕ over ∂B, we try the representation (2.4.1) for
ϕ in $B_i + \partial B$, which immediately yields the boundary relation

$$q \in \int_{\partial B} g(\mathbf{p}, \mathbf{q})\,\sigma(\mathbf{q})\,dq = \phi(\mathbf{p}); \qquad \mathbf{p} \in \partial B. \tag{2.5.1}$$

This is a Fredholm integral equation of the first kind for σ, of which the special
case $\phi = 1$ has already been examined. Little analytical progress can be made
with (2.5.1) directly. However, since ϕ exists everywhere in B_i, it follows that
ϕ_i' exists over ∂B and hence that σ also satisfies equation (2.4.2). This has
the general solution (2.4.4) generating the class of potentials (2.4.5). Clearly
any member of (2.4.5) can only differ from ϕ by a constant which we may
eliminate by suitably choosing k. The appropriate value of k defines that
member of (2.4.4) which solves equation (2.5.1), and the solution is unique.
[The representation (2.5.1) may be readily used to prove that the value of a

harmonic function at the centre of a sphere equals its mean value on the surface of that sphere bearing in mind (2.3.11).]

Utilising the representation (2.4.6) for ϕ in $B_e + \partial B$, the exterior Dirichlet problem is also formulated by equation (2.5.1). Operating upon the left-hand side of (2.5.1) with $\int_{\partial B} \dots \lambda(\mathbf{p})\, dp$, where λ satisfies equation (2.3.2), we find

$$\int_{\partial B} \lambda(\mathbf{p}) \left\{ \int_{\partial B} g(\mathbf{p}, \mathbf{q})\, \sigma(\mathbf{q})\, dq \right\} dp = \int_{\partial B} \sigma(\mathbf{q}) \left\{ \int_{\partial B} g(\mathbf{p}, \mathbf{q})\, \lambda(\mathbf{p})\, dp \right\} dq$$

$$= \int_{\partial B} \sigma(\mathbf{q})\, dq \qquad (2.5.2)$$

by an application of Fubini's theorem, whence

$$\int_{\partial B} \sigma(\mathbf{q})\, dq = \int_{\partial B} \phi(\mathbf{p})\, \lambda(\mathbf{p})\, dq. \qquad (2.5.3)$$

This relation provides a useful check on σ and forms the Dirichlet counterpart of the Neumann relation (2.4.11). [By virtue of

$$\sigma = -\frac{1}{4\pi}(\phi_i' + \phi_e') \qquad \text{and} \qquad \lambda = \frac{1}{4\pi} V_e',$$

where V is the external potential generated by λ, we may write (2.5.3) in the form

$$\int_{\partial B} \{\phi V_e' - \phi_e' V\}\, dq = 0$$

since $V = 1$ on ∂B. Also

$$\int_{\partial S} \{\phi V' - \phi' V\}\, dq = O(r^{-1})$$

where ∂S denotes a large spherical surface of radius r enclosing ∂B. Hence, bearing in mind that ϕ, V are harmonic functions in the domain bounded by $\partial B + \partial S$, we may interpret (2.5.3) as an example of Green's reciprocal formula

(3.1.7) since the integral over ∂S necessarily vanishes (cf. footnote on p. 19).]
An immediate deduction from (2.5.3) is that $\phi = O(r^{-2})$ at infinity if it satisfies
the boundary condition

$$\int_{\partial B} \phi(\mathbf{p})\, \lambda(\mathbf{p})\, dp = 0. \tag{2.5.4}$$

Conversely, if $\phi = O(r^{-2})$ at infinity, we may infer that (2.5.4) holds.

Equation (2.5.1) essentially covers the theory of electrostatic condensers.
The condenser is defined mathematically as a closed conducting surface ∂B_0
maintained at zero potential (earthed), which encloses a second closed
conducting surface ∂B_1 maintained at unit potential. These potentials are
generated by continuous source distributions over ∂B_0 and ∂B_1 which satisfy
the boundary relations

$$\int_{\partial B_0} g(\mathbf{p}, \mathbf{q})\, \sigma(\mathbf{q})\, dq + \int_{\partial B_1} g(\mathbf{p}, \mathbf{q})\, \sigma(\mathbf{q})\, dq = \begin{cases} 0; & \mathbf{p} \in \partial B_0, \\ 1; & \mathbf{p} \in \partial B_1. \end{cases} \tag{2.5.5}$$

These constitute a pair of coupled integral equations formally covered by
(2.5.1) if we regard ∂B as consisting of two distinct sheets $\partial B_0, \partial B_1$. An inherent
property of equations (2.5.5) is that

$$\int_{\partial B_0} \sigma(\mathbf{q})\, dq + \int_{\partial B_1} \sigma(\mathbf{q})\, dq = 0. \tag{2.5.6}$$

We prove this by noting that the source distributions generate a potential V
in the infinite domain B_e exterior to ∂B_0, characterised by

(a) $\qquad\qquad V = 0 \quad \text{over} \quad \partial B_0$

(b) $\qquad\qquad V \to \left\{ \int_{\partial B_0} \sigma(\mathbf{q})\, dq + \int_{\partial B_1} \sigma(\mathbf{q})\, dq \right\} r^{-1} \quad \text{as} \quad r \to \infty$

whence $V = 0$ everywhere in B_e. Accordingly $V_e' = 0$ over ∂B_0, so that
$\int_{\partial B_0} V_e'(\mathbf{p})\, dp = 0$, from which (2.5.6) follows by the Gauss flux theorem.

2.6 Dirichlet problems: double-layer formulation

The double-layer representation

$$\phi(\mathbf{p}) = \int_{\partial B} g(\mathbf{p}, \mathbf{q})_i'\, \mu(\mathbf{q})\, dq; \qquad \mathbf{p} \in B_i, \tag{2.6.1}$$

where μ is a source density to be determined, provides a classically preferred alternative to (2.4.1). The integral jumps by $-2\pi\mu(\mathbf{p})$ at $\mathbf{p} \in \partial B$, in accordance with (1.4.6), whereas ϕ remains continuous at ∂B, so yielding the boundary relation

$$\phi(\mathbf{p}) = \int_{\partial B} g(\mathbf{p}, \mathbf{q})_i'\, \mu(\mathbf{q})\, dq + 2\pi\mu(\mathbf{p}); \qquad \mathbf{p} \in \partial B. \tag{2.6.2}$$

This is a Fredholm integral equation of the second kind for μ in terms of ϕ, with homogeneous component

$$\int_{\partial B} g(\mathbf{p}, \mathbf{q})_i'\, \mu(\mathbf{q})\, dq + 2\pi\mu(\mathbf{p}) = 0; \qquad \mathbf{p} \in \partial B, \tag{2.6.3}$$

which is mathematically equivalent to equation (2.4.9). Since the latter has no non-trivial solution, neither has equation (2.6.3), from which we deduce that equation (2.6.2) has a unique solution for arbitrary continuous values of ϕ given over ∂B. For example putting $\phi = 1$ in (2.6.2), it follows from (1.4.11) that $\mu = 1/4\pi$. Introducing this into (2.6.1) yields $\phi = 1$ in B_i by virtue of (1.4.10).

The representation

$$\phi(\mathbf{p}) = \int_{\partial B} g(\mathbf{p}, \mathbf{q})_e'\, \mu(\mathbf{q})\, dq; \qquad \mathbf{p} \in B_e \tag{2.6.4}$$

does not in general provide an alternative to (2.4.6) because it has $O(r^{-2})$ behaviour at infinity whereas the required ϕ generally has $O(r^{-1})$ behaviour at infinity. We may see this directly by introducing the boundary equation

$$\phi(\mathbf{p}) = \int_{\partial B} g(\mathbf{p}, \mathbf{q})_e'\, \mu(\mathbf{q})\, dq + 2\pi\mu(\mathbf{p}); \qquad \mathbf{p} \in \partial B \tag{2.6.5}$$

corresponding to (2.6.2). The homogeneous form of this equation is mathematically equivalent to (2.3.5), and so it admits the non-trivial solution $\mu = 1$, with a corresponding non-trivial solution of the transpose equation (2.3.4). Therefore equation (2.6.5) only admits a solution if $\int_{\partial B} \phi(\mathbf{p}) \, \lambda(\mathbf{p}) \, \mathrm{d}p = 0$, which is the condition (2.5.4) for $O(r^{-2})$ behaviour at infinity. This limitation of the representation (2.6.4) could be surmounted by writing

$$\phi(\mathbf{p}) = \int_{\partial B} g(\mathbf{p}, \mathbf{q})'_e \, \mu(\mathbf{q}) \, \mathrm{d}q + \frac{c}{|\mathbf{p}|}; \qquad \mathbf{p} \in B_e, \tag{2.6.6}$$

i.e.

$$\phi(\mathbf{p}) - \frac{c}{|\mathbf{p}|} = \int_{\partial B} g(\mathbf{p}, \mathbf{q})'_e \, \mu(\mathbf{q}) \, \mathrm{d}q; \qquad \mathbf{p} \in B_e, \tag{2.6.7}$$

where c is a constant chosen so that

$$\int_{\partial B} \lambda(\mathbf{p}) \, \phi(\mathbf{p}) \, \mathrm{d}p - c \int_{\partial B} \frac{\lambda(\mathbf{p})}{|\mathbf{p}|} \, \mathrm{d}p = 0. \tag{2.6.8}$$

We then replace $\phi(\mathbf{p})$ by $\phi(\mathbf{p}) - (c/|\mathbf{p}|)$ in (2.6.5) to compute a solution which defines the representation (2.6.6).

3

Green's Formula

3.1 Introduction

We now examine a formulation of boundary-value problems in which the boundary data play the role of source densities. Thus suppose there exists a harmonic function in B_i which assumes a continuous set of boundary values $\phi(\mathbf{q})$, and boundary normal derivatives $\phi'_i(\mathbf{q})$, at $\mathbf{q} \in \partial B$. Regarded as source densities, these generate the double-layer potential

$$\int_{\partial B} g(\mathbf{p}, \mathbf{q})'_i \phi(\mathbf{q}) \, d\mathbf{q}; \qquad \mathbf{p} \in B_i \qquad (3.1.1)$$

and the simple-layer potential

$$\int_{\partial B} g(\mathbf{p}, \mathbf{q}) \phi'_i(\mathbf{q}) \, d\mathbf{q}; \qquad \mathbf{p} \in B_i \qquad (3.1.2)$$

respectively. According to Green's formula the superposition of these two potentials yields ϕ in B_i, viz.

$$\int_{\partial B} g(\mathbf{p}, \mathbf{q})'_i \phi(\mathbf{q}) \, dq - \int_{\partial B} g(\mathbf{p}, \mathbf{q}) \phi'_i(\mathbf{q}) \, dq = 4\pi \phi(\mathbf{p}); \qquad \mathbf{p} \in B_i, \qquad (3.1.3)$$

so providing a fundamental link between the theory of harmonic functions and potential theory.

It would be possible to prove (3.1.3) directly by transforming the surface integrals (3.1.1), (3.1.2) into volume integrals using the divergence theorem.

34

A more general approach is to start with the identity

$$\int_{\partial B} [\phi\psi_i' - \psi\phi_i'] \, dq = - \int_{B_i} [\phi\nabla^2\psi - \psi\nabla^2\phi] \, dQ, \qquad (3.1.4)$$

which holds for a wide class of scalar potential fields ϕ, ψ in $B_i + \partial B$, including Newtonian potentials generated by discrete sources in B_i. We prove this by noting from the divergence theorem that

$$\int_{\partial B} \phi\psi_i' \, dq = \int_{\partial B} \phi\nabla\psi \cdot \mathbf{n}_i \, dq = - \int_{B_i} [\phi\nabla^2\psi + \nabla\phi \cdot \nabla\psi] \, dQ, \quad (3.1.5)$$

$$\int_{\partial B} \psi\phi_i' \, dq = \int_{\partial B} \psi\nabla\phi \cdot \mathbf{n}_i \, dq = - \int_{B_i} [\psi\nabla^2\phi + \nabla\psi \cdot \nabla\phi] \, dQ, \quad (3.1.6)$$

whence (3.1.4) follows by subtraction. If ϕ, ψ are harmonic functions of $\mathbf{q}$ in B_i, then

$$\nabla^2\phi = 0, \qquad \nabla^2\psi = 0$$

and the right-hand side of (3.1.4) vanishes, so yielding Green's reciprocal formula

$$\int_{\partial B} \lceil \phi\psi_i' - \psi\phi_i' \rceil \, dq = 0. \qquad (3.1.7)$$

On the other hand, if ϕ is a harmonic function in B_i and

$$\psi(\mathbf{q}) = g(\mathbf{q}, \mathbf{p}); \quad \mathbf{q} \in B_i, \qquad (3.1.8)$$

then

$$\nabla^2\phi = 0, \qquad \nabla^2\psi(\mathbf{q}) = -4\pi\delta(\mathbf{q}, \mathbf{p}); \qquad \mathbf{q} \in B_i. \qquad (3.1.9)$$

Accordingly, the right-hand side of (3.1.4) becomes

$$4\pi \int_{B_i} \phi(\mathbf{q})\delta(\mathbf{q}, \mathbf{p}) \, dQ = 4\pi\phi(\mathbf{p}) \quad \text{if} \quad \mathbf{p} \in B_i, \qquad (3.1.10)$$

and (3.1.3) immediately follows.

Putting $\psi = \phi$ in (3.1.5), we find

$$\int_{\partial B} \phi \phi'_i \, dq = -\int_{B_i} |\nabla \phi|^2 \, dQ, \qquad\qquad (3.1.11)$$

which relation may be utilised to prove the Dirichlet, Neumann and Robin uniqueness theorems mentioned in Chapter 2 (Duff, 1956).

The simple-layer potential (3.1.2) remains continuous at $\mathbf{p} \in \partial B$, but the double-layer potential (3.1.1) jumps by an amount $-2\pi\phi(\mathbf{p})$, so yielding the boundary formula

$$\int_{\partial B} g(\mathbf{p}, \mathbf{q})'_i \phi(\mathbf{q}) \, dq - \int_{\partial B} g(\mathbf{p}, \mathbf{q}) \phi'_i(\mathbf{q}) \, dq = 2\pi\phi(\mathbf{p}); \qquad \mathbf{p} \in \partial B. \quad (3.1.12)$$

This also follows by introducing

$$4\pi \int_{B_i} \phi(\mathbf{q}) \delta(\mathbf{q}, \mathbf{p}) \, dQ = 2\pi\phi(\mathbf{p}); \quad \mathbf{p} \in \partial B, \qquad\qquad (3.1.13)$$

in place of (3.1.10), into the right-hand side of (3.1.4). Formula (3.1.12) provides a functional constraint between ϕ and ϕ'_i over ∂B which ensures their compatibility as boundary data. If, for instance, values of ϕ and ϕ'_i are separately prescribed over ∂B, they do not appertain to the same harmonic function in $B_i + \partial B$ unless (3.1.12) is satisfied.

As $\mathbf{p}$ crosses ∂B into B_e, the double-layer potential again jumps by $-2\pi\phi(\mathbf{p})$, so yielding the identity

$$\int_{\partial B} g(\mathbf{p}, \mathbf{q})'_i \phi(\mathbf{q}) \, dq - \int_{\partial B} g(\mathbf{p}, \mathbf{q}) \phi'_i(\mathbf{q}) \, dq = 0; \qquad \mathbf{p} \in B_e. \quad (3.1.14)$$

This holds for any $\mathbf{p} \in B_e$, as follows for instance by introducing

$$\int_{B_i} \phi(\mathbf{q}) \delta(\mathbf{q}, \mathbf{p}) \, dQ = 0; \quad \mathbf{p} \in B_e, \qquad\qquad (3.1.15)$$

in place of (3.1.10), into the right-hand side of (3.1.4). Alternatively, (3.1.14) may be viewed as a particular case of (3.1.7) since both $\phi(\mathbf{q})$ and $g(\mathbf{q}, \mathbf{p})$ are harmonic functions of $\mathbf{q}$ in B_i if $\mathbf{p} \in B_e$.

Analogous results hold for a regular exterior harmonic function, viz.

$$\int_{\partial B} g(\mathbf{p}, \mathbf{q})'_e \phi(\mathbf{q}) \, dq - \int_{\partial B} g(\mathbf{p}, \mathbf{q}) \phi'_e(\mathbf{q}) \, dq = 4\pi\phi(\mathbf{p}); \quad \mathbf{p} \in B_e, \qquad (3.1.16)$$

$$\int_{\partial B} g(\mathbf{p}, \mathbf{q})'_e \phi(\mathbf{q}) \, dq - \int_{\partial B} g(\mathbf{p}, \mathbf{q}) \phi'_e(\mathbf{q}) \, dq = 2\pi\phi(\mathbf{p}); \quad \mathbf{p} \in \partial B, \qquad (3.1.17)$$

$$\int_{\partial B} g(\mathbf{p}, \mathbf{q})'_e \phi(\mathbf{q}) \, dq - \int_{\partial B} g(\mathbf{p}, \mathbf{q}) \phi'_e(\mathbf{q}) \, dq = 0; \qquad\qquad \mathbf{p} \in B_i, \qquad (3.1.18)$$

corresponding to (3.1.3), (3.1.12), (3.1.14) respectively. To prove (3.1.16) for instance, we enclose ∂B by a large spherical surface ∂S of radius r centred upon the origin, and we apply (3.1.3) to B_e. The contributions generated from ∂S are

$$\int_{\partial S} g(\mathbf{p}, \mathbf{q})' \phi(\mathbf{q}) \, dq = O(r^{-2}) \cdot O(r^{-1}) \cdot r^2 = O(r^{-1}),$$

$$\int_{\partial S} g(\mathbf{p}, \mathbf{q}) \phi'(\mathbf{q}) \, dq = O(r^{-1}) \cdot O(r^{-2}) \cdot r^2 = O(r^{-1}),$$

as $r \to \infty$, showing that the only contributions which count are those arising from ∂B (see footnote on p. 19 Chapter 2). A similar argument shows that the result analogous to (3.1.11) is

$$\int_{\partial B} \phi\phi'_e \, dq = -\int_{B_e} |\nabla\phi|^2 \, dQ. \qquad (3.1.19)$$

3.2 Green's boundary formula: interior problems

Green's formula generates ϕ throughout B_i assuming that ϕ and ϕ'_i are both available and compatible over ∂B. However, according to the fundamental existence theorems, either ϕ alone or ϕ'_i alone over ∂B essentially suffices to determine ϕ throughout B_i. Thus the application of Green's formula requires more boundary information than is necessary from the point of view of the fundamental existence theorems. In so far as these theorems are paralleled by well-posed boundary-value problems, Green's formula cannot be exploited immediately as a means of solving such problems. One well-known approach is to eliminate the redundancy in Green's formula by constructing a Green's function, and this enables progress to be made for specialised boundaries. An alternative strategy is to exploit formula (3.1.12) which couples ϕ and ϕ_i' over ∂B so that they constitute a compatible pair. Combining this global relation with the prescribed local relation (2.1.3), we have sufficient equations to determine ϕ, ϕ_i' separately over ∂B, whence Green's formula generates ϕ throughout B_i.

The latter strategy may be illustrated in detail by reference to the Dirichlet problem. Thus, given ϕ over ∂B, relation (3.1.12) becomes an integral equation of the first kind for ϕ_i' in terms of ϕ, viz.

$$\int_{\partial B} g(\mathbf{p}, \mathbf{q})\phi_i'(\mathbf{q})\,dq = \int_{\partial B} g(\mathbf{p}, \mathbf{q})_i'\phi(\mathbf{q})\,dq - 2\pi\phi(\mathbf{p}); \qquad \mathbf{p} \in \partial B. \quad (3.2.1)$$

This equation is of the type (2.5.1) which has been proved to have a unique solution. However it remains to be proved that the solution satisfies $\int_{\partial B} \phi_i'(\mathbf{q})\,dq = 0$. Operating upon both sides of (3.2.1) with $\int_{\partial B} \ldots \lambda(\mathbf{p})\,dp$, where λ satisfies (2.3.2), the left-hand side becomes

$$\int_{\partial B} \lambda(\mathbf{p})\left\{\int_{\partial B} g(\mathbf{p}, \mathbf{q})\phi_i'(\mathbf{q})\,dq\right\}dp = \int_{\partial B} \phi_i'(\mathbf{q})\left\{\int_{\partial B} \lambda(\mathbf{p})g(\mathbf{p}, \mathbf{q})\,dp\right\}dq$$

$$= \int_{\partial B} \phi_i'(\mathbf{q})\,dq. \quad (3.2.2)$$

The first term on the right-hand side becomes

$$\int_{\partial B} \lambda(\mathbf{p})\left\{\int_{\partial B} g(\mathbf{p}, \mathbf{q})_i'\phi(\mathbf{q})\,dq\right\}dp = \int_{\partial B} \phi(\mathbf{q})\left\{\int_{\partial B} \lambda(\mathbf{p})g(\mathbf{p}, \mathbf{q})_i'\,dp\right\}dq$$

$$= \int_{\partial B} \phi(\mathbf{q})\left\{\int_{\partial B} g_i'(\mathbf{q}, \mathbf{p})\lambda(\mathbf{p})\,dp\right\}dq = 2\pi\int_{\partial B} \phi(\mathbf{q})\lambda(\mathbf{q})\,dq \quad (3.2.3)$$

on bearing in mind (2.3.4) with $\mathbf{p}, \mathbf{q}$ interchanged. Also the second term on the right-hand side becomes

$$-2\pi\int_{\partial B} \phi(\mathbf{p})\lambda(\mathbf{p})\,dp,$$

which cancels out (3.2.3) so proving that $\int_{\partial B} \phi_i'(\mathbf{q})\, dq = 0$ as expected.

The simplest Dirichlet problem is $\phi = 1$ given over ∂B. Putting this condition into the right-hand side of equation (3.2.1) yields

$$\int_{\partial B} g(\mathbf{p}, \mathbf{q})_i'\, dq - 2\pi = 2\pi - 2\pi = 0$$

by virtue of (1.4.11). Hence the equation becomes

$$\int_{\partial B} g(\mathbf{p}, \mathbf{q})\phi_i'(\mathbf{q})\, dq = 0; \quad \mathbf{p} \in \partial B,$$

which implies $\phi_i' = 0$. Introducing the compatible boundary data $\phi = 1$, $\phi_i' = 0$ into (3.1.3), we see without difficulty that they generate $\phi = 1$ throughout B, as expected.

Given ϕ_i' over ∂B (Neumann problem), relation (3.1.12) becomes a generalised Fredholm integral equation of the second kind for ϕ in terms of ϕ_i', viz.

$$\int_{\partial B} g(\mathbf{p}, \mathbf{q})_i'\phi(\mathbf{q})\, dq - 2\pi\phi(\mathbf{p}) = \int_{\partial B} g(\mathbf{p}, \mathbf{q})\phi_i'(\mathbf{q})\, dq; \quad \mathbf{p} \in \partial B. \quad (3.2.4)$$

The homogeneous form of this equation is

$$\int_{\partial B} g(\mathbf{p}, \mathbf{q})_i'\phi(\mathbf{q})\, dq - 2\pi\phi(\mathbf{p}) - 0, \quad (3.2.5)$$

which evidently has the non-trivial solution $\phi = k$ (an arbitrary constant). The corresponding adjoint equation is

$$\int_{\partial B} g_i'(\mathbf{p}, \mathbf{q})\lambda(\mathbf{q})\, dq - 2\pi\lambda(\mathbf{p}) = 0, \quad (3.2.6)$$

which has a non-trivial solution that satisfies (2.3.2). Hence, according to generalised Fredholm theory, equation (3.2.4) has a solution if and only if

$$\int_{\partial B} \lambda(\mathbf{p}) \left\{ \int_{\partial B} g(\mathbf{p}, \mathbf{q})\phi_i'(\mathbf{q})\, dq \right\} dp = 0,$$

i.e.

$$\int_{\partial B} \phi_i'(\mathbf{q}) \left\{ \int_{\partial B} \lambda(\mathbf{p})g(\mathbf{p}, \mathbf{q})\, dp \right\} dq = 0,$$

i.e.

$$\int_{\partial B} \phi'_i(\mathbf{q}) \, dq = 0,$$

which is the expected Gauss condition. Once a particular solution ϕ_0 of equation (3.2.4) has been achieved, the general solution appears as $\phi = \phi_0 + k$.

3.3 Green's boundary formula: exterior problems

The exterior boundary formula yields a corresponding formulation of the exterior Dirichlet problem. Thus, given ϕ over ∂B, it becomes an integral equation of the first kind for ϕ'_e in terms of ϕ, viz.

$$\int_{\partial B} g(\mathbf{p}, \mathbf{q})\phi'_e(\mathbf{q}) \, dq = \int_{\partial B} g(\mathbf{p}, \mathbf{q})'_e \phi(\mathbf{q}) \, dq - 2\pi\phi(\mathbf{p}); \qquad \mathbf{p} \in \partial B. \quad (3.3.1)$$

This has a unique solution, in line with equation (3.2.1), but the solution does not necessarily satisfy $\int_{\partial B} \phi'_e(\mathbf{q}) \, dq = 0$. This is because the result (3.2.3) changes sign on bearing in mind that equation (2.3.4) may be written

$$\int_{\partial B} g'_e(\mathbf{p}, \mathbf{q})\lambda(\mathbf{q}) \, dq + 2\pi\lambda(\mathbf{p}) = 0; \qquad \mathbf{p} \in \partial B. \quad (3.3.2)$$

We deduce that

$$\int \phi'_e(\mathbf{q}) \, dq = -4\pi \int \phi(\mathbf{p})\lambda(\mathbf{p}) \, dp, \quad (3.3.3)$$

a conclusion which also follows by eliminating $\int_{\partial B} \sigma(\mathbf{q}) \, dq$ between (2.4.11) and (2.5.3).

Putting $\phi = 1$ into the right-hand side of equation (3.3.1) yields

$$\int_{\partial B} g(\mathbf{p}, \mathbf{q})\phi'_e(\mathbf{q}) \, dq = -2\pi - 2\pi = -4\pi; \qquad \mathbf{p} \in \partial B \quad (3.3.4)$$

by virtue of (1.4.15), whence $\phi'_e = -4\pi\lambda$ by comparison with (2.3.2). This means that ϕ'_e is proportional to the charge in the capacitance problem of electrostatics. We may therefore solve the latter problem by first constructing ϕ in B_e subject to

$$\phi = 1 \quad \text{over} \quad \partial B, \qquad \phi = O(r^{-1}) \quad \text{as} \quad r \to \infty$$

and then computing ϕ'_e over ∂B. This is the classical approach (Jeans, 1920).

Given ϕ'_e over ∂B (exterior Neumann problem), relation (3.1.17) becomes a generalised Fredholm integral equation of the second kind for ϕ in terms of ϕ'_e, viz.

$$\int_{\partial B} g(\mathbf{p}, \mathbf{q})'_e \phi(\mathbf{q}) \, dq - 2\pi\phi(\mathbf{p}) = \int_{\partial B} g(\mathbf{p}, \mathbf{q})\phi'_e(\mathbf{q}) \, dq; \qquad \mathbf{p} \in \partial B. \quad (3.3.5)$$

In this case the adjoint homogeneous equation is mathematically equivalent to (2.4.8), which has no non-trivial solution. Consequently equation (3.3.5) has a unique solution without any restriction on ϕ'_e. This analysis provides an alternative, extremely useful, formulation of the theory of potential fluid motion perturbed by a fixed rigid obstacle. We merely introduce ϕ'_e as defined by (2.4.12) into the right-hand side of equation (3.3.5), and the solution immediately defines ϕ over ∂B.

3.4 Arbitrariness in Green's formula

According to the theory of this chapter, a given harmonic function ϕ may be represented throughout B_i by Green's formula. On the other hand, according to the theory of Chapter 2, ϕ may also be represented by a simple-layer potential or by a double-layer potential. To bring out the connections between these different representations, we introduce an arbitrary regular exterior harmonic function f into B_e and note that its boundary data satisfy the identity (3.1.18), i.e.

$$\int_{\partial B} g(\mathbf{p}, \mathbf{q})'_e f(\mathbf{q}) \, dq - \int_{\partial B} g(\mathbf{p}, \mathbf{q})f'_e(\mathbf{q}) \, dq = 0; \quad \mathbf{p} \in B_i. \quad (3.4.1)$$

Superposing this on formula (3.1.3) yields the more general continuation formula

$$\int_{\partial B} g(\mathbf{p}, \mathbf{q})'_i [\phi(\mathbf{q}) - f(\mathbf{q})] \, dq - \int_{\partial B} g(\mathbf{p}, \mathbf{q}) \, [\phi'_i(\mathbf{q}) + f'_e(\mathbf{q})] \, dq$$

$$= 4\pi \phi(\mathbf{p}); \qquad \mathbf{p} \in B_i \qquad (3.4.2)$$

already noted by Lamb (1945) and Harrington *et al.* (1969).

We now consider two distinct possibilities for f.

The first is $f = \phi$ over ∂B, providing the representation

$$-\frac{1}{4\pi} \int_{\partial B} g(\mathbf{p}, \mathbf{q}) \, [\phi'_i(\mathbf{q}) + f'_e(\mathbf{q})] \, dq = \phi(\mathbf{p}); \qquad \mathbf{p} \in B_i, \qquad (3.4.3)$$

which may be identified as a simple-layer potential generated by the source density

$$\sigma = -\frac{1}{4\pi} (\phi'_i + f'_e). \qquad (3.4.4)$$

This possibility hinges upon the existence of a unique regular f in B_e satisfying $f = \phi$ over ∂B, as is in fact ensured by the exterior Dirichlet existence theorem. The second possibility is $f'_e = -\phi'_i$ over ∂B, providing the representation

$$\frac{1}{4\pi} \int_{\partial B} g(\mathbf{p}, \mathbf{q})'_i [\phi(\mathbf{q}) - f(\mathbf{q})] \, dq = \phi(\mathbf{p}); \qquad \mathbf{p} \in B_i, \qquad (3.4.5)$$

which may be identified as a double-layer potential generated by the source density

$$\mu = \frac{1}{4\pi} (\phi - f). \qquad (3.4.6)$$

This possibility hinges upon the existence of a unique regular f in B_e satisfying $f'_e = -\phi'_i$ over ∂B, as is in fact ensured by the exterior Neumann existence theorem.

As regards the formulation of Dirichlet problems, that based upon (3.4.5) is superior to either (3.1.3) or (3.4.3) in yielding an integral equation of the second kind, but generally it suffers from the fact that μ carries no physical significance. As regards Neumann problems, there is nothing to choose theoretically between the formulations based on (3.1.3) and (3.4.3), though only the latter seems to be classical. Analogous statements hold for exterior problems.

The representation (3.4.3) remains valid at ∂B, so yielding the boundary relation

$$\int_{\partial B} g(\mathbf{p}, \mathbf{q})\sigma(\mathbf{q})\,\mathrm{d}q = \phi(\mathbf{p}); \qquad \mathbf{p} \in \partial B, \tag{3.4.7}$$

where σ is given by (3.4.4). This may be regarded as an integral equation for σ in terms of ϕ, to which a unique solution exists since ϕ'_i, f'_e uniquely exist. Accordingly we have recovered the formulation (2.5.1) of Dirichlet problems by an adaptation of Green's formula. Similarly the representation (3.4.5) jumps by $-2\pi\mu(\mathbf{p})$ at ∂B, so yielding the boundary relation

$$\int_{\partial B} g(\mathbf{p}, \mathbf{q})'_i\mu(\mathbf{q})\,\mathrm{d}q + 2\pi\mu(\mathbf{p}) = \phi(\mathbf{p}); \qquad \mathbf{p} \in \partial B, \tag{3.4.8}$$

where μ is given by (3.4.6). This may be regarded as an integral equation for μ in terms of ϕ, to which a unique solution exists since f uniquely exists. Accordingly we have recovered the formulation (2.6.2) of Dirichlet problems by an alternative adaptation of Green's formula.

A further interesting application of (3.4.2) is to provide a new derivation of Poisson's integral (see Appendix 4).

3.5 Non-uniqueness of solutions

Our considerations have so far referred to simply-connected domains. However the possibility of a doubly-connected domain, e.g. a torus, is particularly relevant in applications of the theory to potential fluid mechanics. Thus, let us introduce a rigid torus-shaped body B, bounded by a Liapunov surface ∂B, which remains fixed within an infinite ideal fluid. We may imagine a double-layer sheet of uniform source density μ spread across the central opening of the torus, and contained by a vortex line L which lies entirely within the torus (Fig. 3.5.1). This generates a potential field W with properties sufficiently described in Chapter 1. Generally speaking, $\partial W/\partial n_e \neq 0$ on ∂B and therefore W could not be a velocity potential in this case. But we may

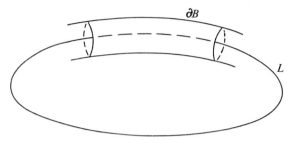

FIG. 3.5.1

introduce a "perturbation" potential ϕ, continuous outside ∂B and regular at infinity, which is uniquely defined by the relation

$$\frac{\partial \phi}{\partial n_e} + \frac{\partial W}{\partial n_e} = 0 \quad \text{on} \quad \partial B, \tag{3.5.1}$$

so yielding an admissible velocity potential $W + \phi$ for any choice of μ. In so far as this potential may be superposed upon a velocity potential transmitted from infinity, the solution of the exterior Neumann problem is not unique for a doubly-connected domain. It may be remarked that

$$\int_{\partial B} \frac{\partial W}{\partial n_e} \, \mathrm{d}q = 0$$

since $W = O(r^{-2})$ as $r \to \infty$. An immediate inference is that $\phi = O(r^{-2})$ as $r \to \infty$ as expected.

Non-uniqueness may also arise from the presence of cusps or sharp edges on ∂B, even for a simply-connected domain. This was first pointed out by Lebesgue (1913, 1924) in connection with the Dirichlet problem, but it also holds for the Neumann problem. We infer that the perturbation of uniform potential flow by a body with a sharp trailing edge cannot be uniquely determined. Aerodynamic theory is largely concerned with the perturbation of uniform airflow by a finite or infinite wing of aerofoil section. The viscosity inherent in airflow greatly complicates the problem as compared with potential flow, but it also enables a unique solution to be achieved on the basis of three physical approximations (Thwaites, 1960):

(a) viscous forces effectively operate only within a thin "boundary layer" enclosing ∂B, outside which potential flow predominates;

(b) these forces create vortex lines which run through the wing in such a way as to produce a circulation around the wing and therefore—according to fundamental hydrodynamic theory—a lift upon the wing;

(c) the distribution of vortex lines and their strengths ensure a zero velocity at the trailing edge (Kutta–Joukowski hypothesis).

These approximations imply that potential flow with suitable vorticity simulates some of the important aerodynamic properties of real airflow.

Considerable mathematical simplifications occur for an infinite two-dimensional wing. We envisage a single vortex line running through the wing parallel to the trailing edge, which line may be regarded as the boundary of a double-layer sheet extending to infinity. This generates a multi-valued velocity potential $\kappa\theta$, with an accompanying circulation $2\pi\kappa$ where κ is a parameter to be determined (Fig. 3.5.2). If the wing perturbs an otherwise uniform velocity potential ψ, the total velocity potential is $\psi + \kappa\theta + \phi$ where ϕ can be uniquely determined for a given κ from the conditions:

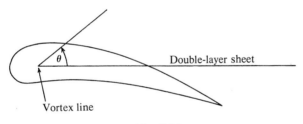

Fig. 3.5.2

(a) ϕ is a single-valued harmonic function everywhere in the fluid;

(b) $\phi = O(1)$ as $r \to \infty$;

(c) $\dfrac{\partial\psi}{\partial n_e} + \kappa\dfrac{\partial\theta}{\partial n_e} + \dfrac{\partial\phi}{\partial n_e} = 0$ over ∂B.

Condition (c) implies that $\phi = O(r^{-1})$ as $r \to \infty$, since

$$\int_{\partial B}\frac{\partial\psi}{\partial n_e}\,dq = 0, \qquad \int_{\partial B}\frac{\partial\theta}{\partial n_e}\,dq = 0. \tag{3.5.2}$$

Once ϕ has been determined as a function of κ, we choose κ so that $\nabla(\psi + \kappa\theta + \phi)$ tends to zero at the trailing edge.

3.6 Liouville's theorem

A function ϕ which is harmonic everywhere and regular at infinity must be identically zero everywhere. To prove this we note that ϕ is harmonic within any finite closed surface ∂B, and it therefore satisfies the interior boundary formula

$$\int_{\partial B} g(\mathbf{p}, \mathbf{q})_i' \phi(\mathbf{q}) \, dq - \int_{\partial B} g(\mathbf{p}, \mathbf{q}) \phi_i'(\mathbf{q}) \, dq = 2\pi\phi(\mathbf{p}); \qquad \mathbf{p} \in \partial B. \quad (3.6.1)$$

It also satisfies the exterior boundary relation

$$\int_{\partial B} g(\mathbf{p}, \mathbf{q})_e' \phi(\mathbf{q}) \, dq - \int_{\partial B} g(\mathbf{p}, \mathbf{q}) \phi_e'(\mathbf{q}) \, dq = 2\pi\phi(\mathbf{p}); \qquad \mathbf{p} \in \partial B. \quad (3.6.2)$$

Now ϕ has the same value in (3.6.1) and (3.6.2) at $\mathbf{q} \in \partial B$ since ϕ is continuous at ∂B, and $\phi_i' + \phi_e' = 0$ at $\mathbf{q} \in \partial B$ since $\partial\phi/\partial n$ is continuous at ∂B. Therefore, superposing (3.6.1) and (3.6.2), we find

$$0 = 4\pi\phi(\mathbf{p}); \qquad \mathbf{p} \in \partial B, \quad (3.6.3)$$

from which the opening theorem follows.

A harmonic function h which is bounded everywhere has the form $h = c + \phi$, where c is a constant and ϕ is a regular harmonic function. Accordingly $\phi = 0$ everywhere and therefore $h = c$ everywhere. This is Liouville's theorem.

4

Two-Dimensional Potential Theory

4.1 The logarithmic potential

All the preceding formulae may be adapted to two dimensions by writing $g(\mathbf{p}, \mathbf{q}) = \log|\mathbf{p} - \mathbf{q}|$ in place of $g(\mathbf{p}, \mathbf{q}) = |\mathbf{p} - \mathbf{q}|^{-1}$ and $-\pi/2$ in place of π. Thus a continuous distribution of simple sources extending over a simple closed Liapunov curve ∂B and of density $\sigma(\mathbf{q})$ at $\mathbf{q} \in \partial B$, generates the simple-layer logarithmic potentials

$$V(\mathbf{p}) = \int_{\partial B} \log|\mathbf{p} - \mathbf{q}|\sigma(\mathbf{q})\,dq; \qquad \mathbf{p} \in B_i, \qquad (4.1.1)$$

$$V(\mathbf{p}) = \int_{\partial B} \log|\mathbf{p} - \mathbf{q}|\sigma(\mathbf{q})\,dq; \qquad \mathbf{p} \in B_e. \qquad (4.1.2)$$

These define harmonic functions in B_i, B_e respectively (Evans, 1927), and they remain continuous at ∂B, and

$$V(\mathbf{p}) = \log|\mathbf{p}|\int_{\partial B} \sigma(\mathbf{q})\,dq - |\mathbf{p}|^{-2}\int_{\partial B} (\mathbf{p}\cdot\mathbf{q})\sigma(\mathbf{q})\,dq + O(|\mathbf{p}|^{-2})$$

$$\text{as} \quad |\mathbf{p}| \to \infty. \qquad (4.1.3)$$

The tangential derivatives of V exist and are continuous at $\mathbf{p} \in \partial B$ provided that σ is Hölder continuous at $\mathbf{p}$, but the normal derivatives are discontinuous.

We write

$$
\left.
\begin{aligned}
\frac{\partial}{\partial n_i} \log|\mathbf{p} - \mathbf{q}| &\equiv \log_i'|\mathbf{p} - \mathbf{q}| \equiv \log|\mathbf{q} - \mathbf{p}|_i' \\[2mm]
\frac{\partial}{\partial n_e} \log|\mathbf{p} - \mathbf{q}| &\equiv \log_e'|\mathbf{p} - \mathbf{q}| \equiv \log|\mathbf{q} - \mathbf{p}|_e'
\end{aligned}
\right\}
\tag{4.1.4}
$$

for the interior and exterior normal derivatives of $\log|\mathbf{p} - \mathbf{q}|$ at $\mathbf{p}$ keeping $\mathbf{q}$ fixed. These have equal status (Chapter 1) and are connected by

$$
\log_i'|\mathbf{p} - \mathbf{q}| + \log_e'|\mathbf{p} - \mathbf{q}| = 0.
\tag{4.1.5}
$$

Utilising this symbolism, the results corresponding to (1.3.7), (1.3.8) are

$$
V_i'(\mathbf{p}) = \int_{\partial B} \log_i'|\mathbf{p} - \mathbf{q}|\,\sigma(\mathbf{q})\,\mathrm{d}q + \pi\sigma(\mathbf{p}); \qquad \mathbf{p} \in \partial B,
\tag{4.1.6}
$$

$$
V_e'(\mathbf{p}) = \int_{\partial B} \log_e'|\mathbf{p} - \mathbf{q}|\,\sigma(\mathbf{q})\,\mathrm{d}q + \pi\sigma(\mathbf{p}); \qquad \mathbf{p} \in \partial B,
\tag{4.1.7}
$$

from which it immediately follows that

$$
V_i'(\mathbf{p}) + V_e'(\mathbf{p}) = 2\pi\sigma(\mathbf{p}); \qquad \mathbf{p} \in \partial B.
\tag{4.1.8}
$$

Also, adapting the analysis of Section 1.5 and anticipating result (4.1.19) below, we find that

$$
\int_{\partial B} V_i'(\mathbf{p})\,\mathrm{d}p = 0, \qquad \int_{\partial B} V_e'(\mathbf{p})\,\mathrm{d}p = 2\pi \int_{\partial B} \sigma(\mathbf{p})\,\mathrm{d}p.
\tag{4.1.9}
$$

As a simple example, take ∂B to be a circle of radius a, with $\sigma = \sigma_0$ (a constant) over ∂B, in which case V is radially symmetric everywhere. Then, since only the first term of (4.1.3) survives,

$$
V(\mathbf{p}) = \log|\mathbf{p}| \int_{\partial B} \sigma(\mathbf{q})\,\mathrm{d}q; \qquad \mathbf{p} \in B_e,
$$

i.e. $V(r) = 2\pi a\sigma_0 \log r; \qquad r \geqslant a,$

using a more familiar notation. Also

$$V(r) = V(a) = 2\pi a \sigma_0 \log a; \qquad r \leqslant a.$$

Consequently

$$V'_e(a) = \left(\frac{dV}{dr}\right)_{r=a} = 2\pi\sigma_0, \quad V'_i(a) = \left(-\frac{dV}{dr}\right)_{r=a} = 0$$

yielding

$$V'_e(a) + V'_i(a) = 2\pi\sigma_0$$

as expected. More general considerations for the circle are discussed in Appendix 5.

A continuous distribution of double sources over ∂B, of density $\mu(\mathbf{q})$ at $\mathbf{q} \in \partial B$, generates the double-layer logarithmic potentials

$$W(\mathbf{p}) = \int_{\partial B} \log|\mathbf{p} - \mathbf{q}|'_i \, \mu(\mathbf{q}) \, dq; \qquad \mathbf{p} \in B_i, \tag{4.1.10}$$

$$W(\mathbf{p}) = \int_{\partial B} \log|\mathbf{p} - \mathbf{q}|'_i \, \mu(\mathbf{q}) \, dq; \qquad \mathbf{p} \in B_e. \tag{4.1.11}$$

These are harmonic functions in B_i, B_e respectively, and

$$W(\mathbf{p}) = O(|\mathbf{p}|^{-2}) \quad \text{as} \quad |\mathbf{p}| \to \infty. \tag{4.1.12}$$

The integral suffers a discontinuity at ∂B defined by

$$\lim_{\mathbf{p}_i \to \mathbf{p}} W(\mathbf{p}_i) = W(\mathbf{p}) - \pi\mu(\mathbf{p}); \qquad \mathbf{p} \in \partial B, \tag{4.1.13}$$

$$\lim_{\mathbf{p}_e \to \mathbf{p}} W(\mathbf{p}_e) = W(\mathbf{p}) + \pi\mu(\mathbf{p}); \qquad \mathbf{p} \in \partial B, \tag{4.1.14}$$

where $\mathbf{p}_i$, $\mathbf{p}_e$ are points on the n_i, n_e—normals emanating from $\mathbf{p} \in \partial B$. As regards normal derivatives, the continuity of μ ensures that

$$\lim_{\mathbf{p}_i \to \mathbf{p}} \frac{\partial W(\mathbf{p}_i)}{\partial n_i} + \lim_{\mathbf{p}_e \to \mathbf{p}} \frac{\partial W(\mathbf{p}_e)}{\partial n_e} = 0; \quad \mathbf{p} \in \partial B. \tag{4.1.15}$$

However, the separate limits necessarily exist only if $\mu \in C^{(2)}$, or μ is at least Hölder continuously differentiable, in the neighbourhood of $\mathbf{p}$. As regards tangential derivatives, the formulae (1.4.9) apply with 2π replaced by $-\pi$.

Putting $\mu = 1$ in (4.1.10) gives, by virtue of the two-dimensional Gauss flux theorem,

$$W(\mathbf{p}) = \int_{\partial B} \log |\mathbf{p} - \mathbf{q}|'_i \, dq = -2\pi; \quad \mathbf{p} \in B_i, \tag{4.1.16}$$

alternatively written

$$W(\mathbf{p}) = \int_{\partial B} \log'_i |\mathbf{q} - \mathbf{p}| \, dq = -2\pi; \quad \mathbf{p} \in B_i. \tag{4.1.17}$$

The latter equality can be proved directly by noting that $\log |\mathbf{q} - \mathbf{p}|$, $\theta(\mathbf{q} - \mathbf{p})$ are conjugate harmonic functions of $\mathbf{q}$ keeping $\mathbf{p}$ fixed (Fig. 4.1.1):

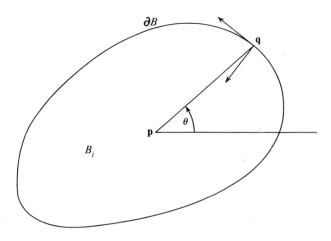

FIG. 4.1.1

these satisfy the Cauchy–Riemann equation

$$\log_i' |\mathbf{q} - \mathbf{p}| = -\frac{\partial \theta (\mathbf{q} - \mathbf{p})}{\partial q}; \qquad \mathbf{q} \in \partial B, \tag{4.1.18}$$

where $\partial / \partial q$ denotes the tangential derivative directed so as to keep B_i on the left, thus yielding

$$\int_{\partial B} \log_i' |\mathbf{q} - \mathbf{p}| \, dq = -\int_{\partial B} \frac{\partial \theta (\mathbf{q} - \mathbf{p})}{\partial q} \, dq = -2\pi; \qquad \mathbf{p} \in B_i.$$

By virtue of (4.1.13), or alternatively from (4.1.18) with $\mathbf{p} \in \partial B$, it follows that

$$\int_{\partial B} \log |\mathbf{p} - \mathbf{q}|_i' \, dq = -\pi; \qquad \mathbf{p} \in \partial B. \tag{4.1.19}$$

More generally, relaxing the restriction to a Liapunov curve, if $\mathbf{p}$ is located at a corner of the boundary having an interior angle $\Omega(\mathbf{p})$, then

$$\int_{\partial B} \log |\mathbf{p} - \mathbf{q}|_i' \, dq = -\Omega(\mathbf{p}); \qquad \mathbf{p} \in \partial B. \tag{4.1.20}$$

By a further application of (4.1.18), it follows that

$$\int_{\partial B} \log |\mathbf{p} - \mathbf{q}|_i' \, dq = 0; \qquad \mathbf{p} \in B_e. \tag{4.1.21}$$

Exterior formulae corresponding to (4.1.16), (4.1.19), (4.1.21) are respectively

$$\int_{\partial B} \log |\mathbf{p} - \mathbf{q}|_e' \, dq = 2\pi; \qquad \mathbf{p} \in B_i, \tag{4.1.22}$$

$$\int_{\partial B} \log |\mathbf{p} - \mathbf{q}|_e' \, dq = \pi; \qquad \mathbf{p} \in \partial B, \tag{4.1.23}$$

$$\int_{\partial B} \log |\mathbf{p} - \mathbf{q}|_e' \, dq = 0; \qquad \mathbf{p} \in B_e, \tag{4.1.24}$$

c

which results may also be proved by closing B_e with a large outer circle ∂S. As an exercise on (4.1.23), take ∂B to be a circle of radius a, with

$$\mathbf{p} = (a \cos \alpha, a \sin \alpha); \qquad \mathbf{q} = (r \cos \theta, r \sin \theta)_{r=a}.$$

Then

$$\log |\mathbf{p} - \mathbf{q}| = \tfrac{1}{2}\log [r^2 + a^2 - 2ra \cos (\theta - \alpha)]_{r=a} = \log \left(2a \sin \left| \frac{\theta - \alpha}{2} \right| \right),$$

$$\log |\mathbf{p} - \mathbf{q}|'_e = \frac{d}{dr} \log |\mathbf{p} - \mathbf{q}|_{r=a} = \frac{1}{2a},$$

$$\int_{\partial B} \log |\mathbf{p} - \mathbf{q}|'_e \, dq = \int_0^{2\pi} (1/2a)a \, d\theta = \pi.$$

It will be noted that $\log |\mathbf{p} - \mathbf{q}|'_e$ is independent of $|\mathbf{p} - \mathbf{q}|$, and it therefore equals $1/2a$ when $\mathbf{p} = \mathbf{q}$.

4.2 Γ-Contours

A distinguishing feature of two-dimensional theory is the existence of contours, termed Γ-contours, for which the equation

$$\int_{\partial B} \log |\mathbf{p} - \mathbf{q}| \lambda(\mathbf{q}) \, dq = 1; \qquad \mathbf{p} \in \partial B \tag{4.2.1}$$

does not have a solution (Jaswon, 1963). For example, if ∂B is a circle of radius a, then $\lambda(\mathbf{q}) = \lambda_0$ (a constant) by symmetry, in which case (4.2.1) becomes

$$\lambda_0 \int_{\partial B} \log |\mathbf{p} - \mathbf{q}| \, dq = \lambda_0 2\pi a \log a = 1. \tag{4.2.2}$$

It follows that λ_0 has no finite value when $a = 1$, and that any finite λ_0 provides a non-trivial solution of the homogeneous equation

$$\int_{\partial B} \log |\mathbf{p} - \mathbf{q}| \lambda(\mathbf{q}) \, dq = 0; \qquad \mathbf{p} \in \partial B \tag{4.2.3}$$

in this case. This feature of the unit circle has been noted by Petrovsky (1954).

The unit circle case may be generalised. Given a contour ∂B defined by the Cartesian equation $f(x, y) = 0$, construct the family of similar contours defined by $f(x/m, y/m) = 0$ where $m > 0$ is a continuously varying parameter (Fig. 4.2.1). Then equation (4.2.1) admits a unique solution for every member

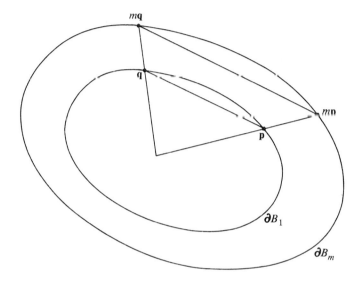

FIG. 4.2.1

of the family with one exception, the Γ-contour, which is characterised by the existence of a non-trivial solution of (4.2.3). To prove this theorem, we label the mth contour ∂B_m, and introduce the normal derivative equation

$$\int_{\partial B_1} \log'_i |\mathbf{p} - \mathbf{q}| \lambda(\mathbf{q}) \, dq + \pi \lambda(\mathbf{p}) = 0; \qquad \mathbf{p} \in \partial B_1 \qquad (4.2.4)$$

appropriate to a uniform potential within the initial contour ∂B_1 ($=\partial B$). Following the analysis of Chapter 2, this equation has a non-trivial solution λ which either uniquely satisfies (4.2.1) or possibly satisfies (4.2.3). In either event we introduce λ into the integral

$$\int_{\partial B_m} \log |\mathbf{p}_m - \mathbf{q}_m| \lambda(\mathbf{q}_m) \, dq_m, \qquad (4.2.5)$$

where $\mathbf{p}_m$, $\mathbf{q}_m$ are points on ∂B_m similarly located to $\mathbf{p}$, $\mathbf{q}$ on ∂B and where $\lambda(\mathbf{q}_m) = \lambda(\mathbf{q})$. Since

$$|\mathbf{p}_m - \mathbf{q}_m| = m|\mathbf{p} - \mathbf{q}|, \qquad dq_m = m\,dq,$$

this integral equals

$$m \log m \int_{\partial B} \lambda(\mathbf{q})\,dq + m \int_{\partial B} \log|\mathbf{p} - \mathbf{q}|\lambda(\mathbf{q})\,dq. \qquad (4.2.6)$$

If λ satisfies (4.2.1), then (4.2.6) reduces to

$$\kappa m \log m + m; \qquad \kappa = \int_{\partial B} \lambda(\mathbf{q})\,dq > 0, \qquad (4.2.7)$$

showing that $m = \exp(-1/\kappa)$ characterises the Γ-contour for this family. If λ satisfies (4.2.3), i.e. if $\partial B \equiv \partial B_1 \equiv \Gamma$, then (4.2.6) reduces to $\kappa m \log m$ showing that $\partial B_m \neq \Gamma$ when $m \neq 1$.

4.3 Two-dimensional Dirichlet problem

We now examine the more general equation

$$\int_{\partial B} \log|\mathbf{p} - \mathbf{q}|\sigma(\mathbf{q})\,dq = \phi(\mathbf{p}); \qquad \mathbf{p} \in \partial B, \qquad (4.3.1)$$

where ϕ is prescribed over ∂B. (Equation (4.3.1) was formulated by Hamel (1949) and also by Volterra (1959), in each case without any further analysis.) Following the analysis of Chapter 2 this equation has a unique solution when $\partial B \neq \Gamma$, and also

$$\int_{\partial B} \phi(\mathbf{p})\lambda(\mathbf{p})\,dp = \int_{\partial B} \lambda(\mathbf{p})\left\{\int_{\partial B} \log|\mathbf{p} - \mathbf{q}|\sigma(\mathbf{q})\,dq\right\}dp,$$

$$= \int_{\partial B} \sigma(\mathbf{q})\left\{\int_{\partial B} \lambda(\mathbf{p})\log|\mathbf{p} - \mathbf{q}|\,dp\right\}dq,$$

$$= \int_{\partial B} \sigma(\mathbf{q}).1.\,dq = \int_{\partial B} \sigma(\mathbf{q})\,dq \qquad (4.3.2)$$

by virtue of (4.2.1). Clearly the interior potential generated by σ solves the interior Dirichlet problem for ∂B. However the exterior potential generated by σ is characterised by logarithmic behaviour at infinity, whereas the classical existence-uniqueness theorem (Kellogg, 1929) specifies $O(1)$ behaviour at infinity. To construct an acceptable exterior harmonic function, we introduce the boundary decomposition

$$\phi = \phi_0 + k \tag{4.3.3}$$

where k is a constant chosen so that

$$\int_{\partial B} \phi(\mathbf{p})\lambda(\mathbf{p})\, dp = k \int_{\partial B} \lambda(\mathbf{p})\, dp \tag{4.3.4}$$

thereby ensuring

$$\int_{\partial B} \phi_0(\mathbf{p})\lambda(\mathbf{p})\, dp = 0. \tag{4.3.5}$$

Corresponding to (4.3.3) we write

$$\sigma = \sigma_0 + k\lambda, \tag{4.3.6}$$

where σ_0 evidently satisfies

$$\int_{\partial B} \log |\mathbf{p} - \mathbf{q}| \sigma_0(\mathbf{q})\, dq = \phi_0(\mathbf{p}); \qquad \mathbf{p} \in \partial B, \tag{4.3.7}$$

$$\int_{\partial B} \sigma_0(\mathbf{q})\, dq = \int_{\partial B} \phi_0(\mathbf{p})\lambda(\mathbf{p})\, dp = 0. \tag{4.3.8}$$

Equation (4.3.7) has a unique solution which generates an exterior potential V_0 coinciding with ϕ_0 on ∂B and characterised by $O(r^{-1})$ behaviour at infinity. Accordingly the decomposition (4.3.3) provides an exterior harmonic function $V_0 + k$ coinciding with ϕ on ∂B and characterised by $O(1)$ behaviour at infinity, thereby solving the exterior Dirichlet problem for ∂B. The same decomposition provides an interior harmonic function $V_0 + k$ identical with that generated directly by σ.

Equation (4.3.1) may not have a solution when $\partial B = \Gamma$, e.g. if $\phi = 1$.

If a solution exists, then in place of (4.3.2) we obtain

$$\int_\Gamma \phi(\mathbf{p})\lambda(\mathbf{p})\,\mathrm{d}p = \int_\Gamma \sigma(\mathbf{q})\cdot 0 \cdot \mathrm{d}q = 0 \tag{4.3.9}$$

since λ now satisfies equation (4.2.3). This is a necessary and sufficient condition for the solution of

$$\int_\Gamma \log|\mathbf{p} - \mathbf{q}|\,\sigma(\mathbf{q})\,\mathrm{d}q = \phi(\mathbf{p}); \qquad \mathbf{p} \in \Gamma \tag{4.3.10}$$

to exist. If (4.3.9) is not satisfied, we introduce the decomposition (4.3.3) which yields

$$\int_\Gamma \log|\mathbf{p} - \mathbf{q}|\,\sigma_0(\mathbf{q})\,\mathrm{d}q = \phi_0(\mathbf{p}); \qquad \mathbf{p} \in \Gamma, \tag{4.3.11}$$

$$\int_\Gamma \phi_0(\mathbf{p})\lambda(\mathbf{p})\,\mathrm{d}p = 0. \tag{4.3.12}$$

There is no difficulty in proving that equation (4.3.11) has a class of solutions of the form $\sigma_0 + m\lambda$ where m is an arbitrary multiplier and $\int_\Gamma \sigma_0(\mathbf{q})\,\mathrm{d}q = 0$. The particular solution σ_0 generates an exterior potential V_0 characterised by $O(r^{-1})$ behaviour at infinity, thereby serving to define an exterior harmonic function $V_0 + k$ which solves the exterior Dirichlet problem for Γ. Similarly we may construct an interior harmonic function $V_0 + k$ which solves the interior Dirichlet problem for Γ. Note that equation (4.3.11) with (4.3.12) covers the case of equation (4.3.10) when (4.3.9) is satisfied.

If a harmonic function ϕ is bounded at infinity and has prescribed values on ∂B which satisfy the condition $\int_{\partial B} \phi(\mathbf{p})\lambda(\mathbf{p})\,\mathrm{d}p = 0$, we may infer that $\phi = O(r^{-1})$ at infinity. Conversely, if $\phi = O(r^{-1})$ at infinity, we may infer that $\int_{\partial B} \phi(\mathbf{p})\lambda(\mathbf{p})\,\mathrm{d}p = 0$. In this case equation (4.3.7), with the side condition (4.3.8), applies *ab initio*, and σ_0 generates a potential V_0 which immediately solves the Dirichlet problem. Parallel statements hold if ϕ satisfies (4.3.9) on Γ.

4.4 Two-dimensional Green's formula

Green's formula (3.1.3), adapted to two dimensions, takes the form

$$\int_{\partial B} \log|\mathbf{p} - \mathbf{q}|'_i \, \phi(\mathbf{q}) \, dq - \int_{\partial B} \log|\mathbf{p} - \mathbf{q}| \, \phi'_i(\mathbf{q}) \, dq = -2\pi\phi(\mathbf{p});$$

$$\mathbf{p} \in B_i. \qquad (4.4.1)$$

For a field point $\mathbf{p} \in \partial B$, located at a corner of the boundary having an internal angle $\Omega(\mathbf{p})$, the right-hand side of (4.4.1) changes to $-\Omega\phi$ bearing in mind the jump from (4.1.17) to (4.1.20), so yielding the boundary formula

$$\int_{\partial B} \log|\mathbf{p} - \mathbf{q}|'_i \, \phi(\mathbf{q}) \, dq - \int_{\partial B} \log|\mathbf{p} - \mathbf{q}| \, \phi'_i(\mathbf{q}) \, dq = -\Omega(\mathbf{p})\phi(\mathbf{p});$$

$$\mathbf{p} \in \partial B. \qquad (4.4.2)$$

In particular, if the boundary is smooth at $\mathbf{p}$, i.e. if $\Omega(\mathbf{p}) = \pi$, (4.4.2) becomes

$$\int_{\partial B} \log|\mathbf{p} - \mathbf{q}|'_i \, \phi(\mathbf{q}) \, dq - \int_{\partial B} \log|\mathbf{p} - \mathbf{q}| \, \phi'_i(\mathbf{q}) \, dq = -\pi\phi(\mathbf{p});$$

$$\mathbf{p} \in \partial B. \qquad (4.4.3)$$

This important identity will be recognised as the real-variable analogue of Plemelj's complex variable formula (Muskhelishvili, 1953a). Given ϕ'_i over ∂B (interior Neumann problem), relation (4.4.3) becomes an integral equation of the second kind for ϕ in terms of ϕ'_i, with analogous properties to equation (3.2.4). Given ϕ over ∂B (interior Dirichlet problem), relation (4.4.3) becomes an integral equation of the first kind for ϕ'_i in terms of ϕ, viz.

$$\int_{\partial B} \log|\mathbf{p} - \mathbf{q}| \, \phi'_i(\mathbf{q}) \, dq = \pi\phi(\mathbf{p}) + \int_{\partial B} \log|\mathbf{p} - \mathbf{q}|'_i \, \phi(\mathbf{q}) \, dq;$$

$$\mathbf{p} \in \partial B. \qquad (4.4.4)$$

This equation falls into the class (4.3.7), since

$$\int_{\partial B} \lambda(\mathbf{p}) \left\{ \pi \phi(\mathbf{p}) + \int_{\partial B} \log |\mathbf{p} - \mathbf{q}|'_i \phi(\mathbf{q}) \, dq \right\} dp$$

$$= \pi \int_{\partial B} \lambda(\mathbf{p}) \phi(\mathbf{p}) \, dp + \int_{\partial B} \phi(\mathbf{q}) \left\{ \int_{\partial B} \log |\mathbf{p} - \mathbf{q}|'_i \lambda(\mathbf{p}) \, dp \right\} dq$$

$$= \pi \int_{\partial B} \lambda(\mathbf{p}) \phi(\mathbf{p}) \, dp - \pi \int_{\partial B} \phi(\mathbf{q}) \lambda(\mathbf{q}) \, dq = 0 \qquad (4.4.5)$$

on bearing in mind (4.2.4) with $\mathbf{p}, \mathbf{q}$ interchanged. It follows that $\int_{\partial B} \phi'_i(\mathbf{q}) \, dq = 0$ as expected. It also follows that a solution exists when $\partial B = \Gamma$, which solution is not unique but can be chosen to satisfy the condition $\int_\Gamma \phi'_i(\mathbf{q}) \, dq = 0$.

Changing i into e yields the analogous exterior formulae, where Ω now signifies the external angle at $\mathbf{p}$. For instance equation (4.4.4) gets replaced by

$$\int_{\partial B} \log |\mathbf{p} - \mathbf{q}| \phi'_e(\mathbf{q}) \, dq = \pi \phi(\mathbf{p}) + \int_{\partial B} \log |\mathbf{p} - \mathbf{q}|'_e \phi(\mathbf{q}) \, dq \, ; \mathbf{p} \in \partial B, \qquad (4.4.6)$$

which is an integral equation of the first kind for ϕ'_e in terms of ϕ. Unlike (4.4.4), however, this is not necessarily of the form (4.3.7) since

$$\int_{\partial B} \lambda(\mathbf{p}) \left\{ \pi \phi(\mathbf{p}) + \int_{\partial B} \log |\mathbf{p} - \mathbf{q}|'_e \phi(\mathbf{q}) \, dq \right\} dp$$

$$= \pi \int_{\partial B} \lambda(\mathbf{p}) \phi(\mathbf{p}) \, dp + \pi \int_{\partial B} \phi(\mathbf{q}) \lambda(\mathbf{q}) \, dq = 2\pi \int_{\partial B} \lambda(\mathbf{p}) \phi(\mathbf{p}) \, dp.$$

Accordingly, whilst equation (4.4.4) is theoretically superior to (4.3.1) for interior Dirichlet problems in that it always has a solution, equation (4.4.6) has no theoretical advantage over (4.3.1) for exterior Dirichlet problems. It may be remarked that

$$\int_{\partial B} \phi'_e(\mathbf{q}) \, dq = 2\pi \int_{\partial B} \lambda(\mathbf{p}) \phi(\mathbf{p}) \, dp, \qquad (4.4.7)$$

as follows by operating upon the left-hand side of (4.4.6) with $\int_{\partial B} \ldots \lambda(\mathbf{p}) \, dp$. Accordingly the exterior analogue of (4.4.1) defines a potential characterised by the behaviour

$$-2\pi\phi \to r^{-1} \int_{\partial B} \phi(\mathbf{q})\,dq - \log r \int_{\partial B} \phi'_e(\mathbf{q})\,dq \quad \text{as} \quad r \to \infty. \qquad (4.4.8)$$

No contributions to $\phi(\mathbf{p})$ at any finite point $\mathbf{p}$ arise from this asymptotic behaviour (see footnote on p. 19) since

$$\int_{\partial S} \log|\mathbf{p} - \mathbf{q}|' \phi(\mathbf{q})\,dq - \int_{\partial S} \log|\mathbf{p} - \mathbf{q}|\phi'(\mathbf{q})\,dq = O(r^{-1})$$

$$\text{as} \quad r = |\mathbf{q}| \to \infty; \qquad \mathbf{p} \in B_n, \qquad (4.4.9)$$

where ∂S is a large circle of radius r enclosing ∂B and $\phi(\mathbf{q})$ has the form (4.4.8). If $O(1)$ behaviour at infinity is particularly required, we write

$$k + \int_{\partial B} \log|\mathbf{p} - \mathbf{q}|'_e \phi(\mathbf{q})\,dq - \int_{\partial B} \log|\mathbf{p} - \mathbf{q}|\phi'_e(\mathbf{q})\,dq = -2\pi\phi(\mathbf{p});$$

$$\mathbf{p} \in B_e, \qquad (4.4.10)$$

where k is a constant to be determined subject to the side condition $\int_{\partial B} \phi'_e(\mathbf{q})\,dq = 0$. Clearly $k\int_{\partial B} \lambda(\mathbf{p})\,dp = -2\pi\int_{\partial B}\phi(\mathbf{p})\lambda(\mathbf{p})\,dp$. This provides a formulation of exterior Dirichlet and Neumann problems which holds for all contours.

4.5 Conformal mapping

Two dimensional potential theory provides a fresh approach to the problem of conformal mapping. It is convenient to begin with the exterior problem. Thus, let $w(z)$ be the (single-valued) analytic function which maps B_e in the complex z-plane onto $|w| > 1$ in the complex w-plane, so that ∂B corresponds to $|w| = 1$ and $z \to \infty$ corresponds to $w \to \infty$. Then, taking $z = 0$ to be located within the interior domain B_i, we have

$$w(z) = ze^{\phi + i\psi}; \qquad z \in B_e + \partial B, \qquad (4.5.1)$$

where ϕ, ψ are conjugate (single-valued) harmonic functions. These are to be determined subject to $O(1)$ behaviour at infinity and to the boundary condition

$$|w| = |z| e^{\phi} = 1; \qquad z \in \partial B,$$

i.e. $\quad \log|z| + \phi = 0; \qquad z \in \partial B. \qquad (4.5.2)$

An admissible representation for ϕ is

$$\phi(\mathbf{p}) = k + \int_{\partial B} \log|\mathbf{p} - \mathbf{q}| \, \sigma_0(\mathbf{q}) \, dq; \qquad \mathbf{p} \in B_e + \partial B, \quad |\mathbf{p}| = |z|, \qquad (4.5.3)$$

subject to the side condition

$$\int_{\partial B} \sigma_0(\mathbf{q}) \, dq = 0, \qquad (4.5.4)$$

where k is a constant to be determined and σ_0 is a source density to be determined. Then (4.5.2) provides the integral equation

$$\log|\mathbf{p}| + k + \int_{\partial B} \log|\mathbf{p} - \mathbf{q}| \, \sigma_0(\mathbf{q}) \, dq = 0; \qquad \mathbf{p} \in \partial B \qquad (4.5.5)$$

for σ_0, the unknown constant k being balanced by (4.5.4). This constant may be computed directly by operating upon equation (4.5.5) with $\int_{\partial B} \dots \lambda(\mathbf{p}) \, dp$ which gives

$$\int_{\partial B} \log|\mathbf{p}| \, \lambda(\mathbf{p}) \, dp + k \int_{\partial B} \lambda(\mathbf{p}) \, dp$$

$$+ \int_{\partial B} \sigma_0(\mathbf{q}) \left\{ \int_{\partial B} \log|\mathbf{p} - \mathbf{q}| \, \lambda(\mathbf{p}) \, dp \right\} dq = 0, \qquad (4.5.6)$$

i.e. $\quad \displaystyle\int_{\partial B} \log|\mathbf{p}| \, \lambda(\mathbf{p}) \, dp + k\kappa + \int_{\partial B} \sigma_0(\mathbf{q}) \, dq = 0; \qquad \kappa = \int_{\partial B} \lambda(\mathbf{q}) \, dq > 0$

since $\int_{\partial B} \log|\mathbf{p} - \mathbf{q}| \, \lambda(\mathbf{p}) \, dp = 1; \mathbf{p} \in B_i + \partial B$. Now $\int_{\partial B} \log|\mathbf{p}| \, \lambda(\mathbf{p}) \, dp = 1$, as follows by putting $\mathbf{q} = 0$ in the preceding equation. Therefore $k\kappa + 1 = 0$, i.e. $k = -1/\kappa$, bearing in mind (4.5.4).

It is interesting to note that the interior mapping problem is covered by $k = 0$ and omitting the side condition (4.5.4), in which case equation (4.5.5) becomes

$$\log|\mathbf{p}| + \int_{\partial B} \log|\mathbf{p} - \mathbf{q}|\,\sigma_0(\mathbf{q})\,dq = 0; \qquad \mathbf{p} \in \partial B. \tag{4.5.7}$$

This yields the relation

$$\int_{\partial B} \log|\mathbf{p}|\,\lambda(\mathbf{p})\,dp + \int_{\partial B} \sigma_0(\mathbf{q})\left\{\int_{\partial B} \log|\mathbf{p} - \mathbf{q}|\,\lambda(\mathbf{p})\,dp\right\}dq = 0$$

in place of (4.5.6), yielding

$$1 + \int_{\partial B} \sigma_0(\mathbf{q})\,dq = 0 \tag{4.5.8}$$

in place of (4.5.4).

If $\partial B = \Gamma$, then $\int_\Gamma \log|\mathbf{p} - \mathbf{q}|\,\lambda(\mathbf{p})\,dp = 0; \mathbf{q} \in B_i + \Gamma$ showing from (4.5.6) that $k = 0$ even for exterior problems. Equation (4.5.5) now becomes

$$\log|\mathbf{p}| + \int_\Gamma \log|\mathbf{p} - \mathbf{q}|\,\sigma_0(\mathbf{q})\,dq = 0; \qquad \mathbf{p} \in \Gamma, \tag{4.5.9}$$

which has a solution (for both exterior and interior problems) since $\int_\Gamma \log|\mathbf{p}|\,\lambda(\mathbf{p})\,dp = 0$, and this solution can be chosen to satisfy $\int_\Gamma \sigma_0(\mathbf{q})\,dq = 0$.

The transfinite diameter c associated with ∂B may be defined (Landkof, 1972) through the relation

$$w(z) = \frac{z}{c} + O(1) \quad \text{as} \quad z \to \infty. \tag{4.5.10}$$

By comparison with (4.5.1), bearing in mind from (4.5.3), (4.5.4) that $\phi = k + O(r^{-1})$ as $r \to \infty$, we find

$$c^{-1} = e^k = e^{-1/\kappa}; \qquad \partial B \neq \Gamma, \tag{4.5.11}$$

$$c^{-1} = e^0 = 1; \qquad \partial B = \Gamma. \tag{4.5.12}$$

Therefore Γ-contours are characterised by unit transfinite diameters.

With σ_0 known, the exterior conjugate harmonic function ψ immediately appears as

$$\psi(\mathbf{p}) = \psi_0 + \int_{\partial B} \theta(\mathbf{p} - \mathbf{q})\sigma_0(\mathbf{q}) \, dq; \qquad \mathbf{p} \in B_e + \partial B, \qquad (4.5.13)$$

where ψ_0 is an arbitrary constant (producing an allowed rigid-body rotation in the w-plane). Here $\theta(\mathbf{p} - \mathbf{q})\sigma_0(\mathbf{q})$, $\log|\mathbf{p} - \mathbf{q}|\,\sigma_0(\mathbf{q})$ are conjugate harmonic functions of $\mathbf{p}$. Accordingly, θ denotes the angle between a fixed direction through $\mathbf{q}$ and the vector joining $\mathbf{q}$ to $\mathbf{p}$. For further details see Section 11.7. The potential (4.5.13) has the asymptotic behaviour

$$\psi(\mathbf{p}) = \psi_0 + \theta(\mathbf{p}) \int_{\partial B} \sigma_0(\mathbf{q}) \, dq + O(|\mathbf{p}|^{-1}) \quad \text{as} \quad |\mathbf{p}| \to \infty, \qquad (4.5.14)$$

$$\text{i.e.} \quad \psi(\mathbf{p}) = \psi_0 + O(|\mathbf{p}|^{-1}) \quad \text{as} \quad |\mathbf{p}| \to \infty, \qquad (4.5.15)$$

being therefore single-valued in B_e as required. The interior conjugate harmonic function has the form (4.5.13), being inherently single-valued without any restriction on σ_0.

A ring-shaped domain B, bounded externally by ∂B_0 and internally by ∂B_1, may be mapped onto the region between two concentric circles having a definite ratio of their radii (Fig. 4.5.1). Suppose ∂B_0 maps onto the outer circle $|w| = 1$, then ∂B_1 maps onto the inner circle $|w| = \rho < 1$. As before

$$w(z) = ze^{\phi + i\psi}; \qquad z \in B + \partial B_0 + \partial B_1, \qquad (4.5.16)$$

where ϕ, ψ are conjugate (single-valued) harmonic functions to be determined subject to the boundary conditions

$$|w| = |z|e^{\phi} = 1; \qquad z \in \partial B_0,$$

$$|w| = |z|e^{\phi} = \rho; \qquad z \in \partial B_1,$$

$$\left.\begin{array}{ll} \text{i.e.} \quad \log|z| + \phi = 0; & z \in \partial B_0, \\[2mm] \log|z| + \phi = \log \rho; & z \in \partial B_1. \end{array}\right\} \qquad (4.5.17)$$

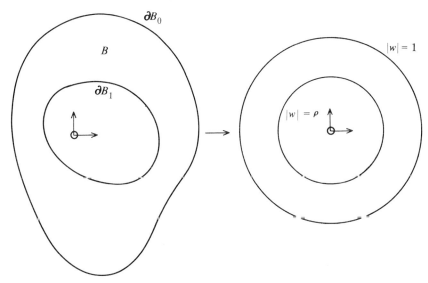

FIG. 4.5.1.

Admissible representations for ϕ, ψ are

$$\phi(\mathbf{p}) = \int_{\partial B_0} \log|\mathbf{p} - \mathbf{q}|\sigma(\mathbf{q})\,dq + \int_{\partial B_1} \log|\mathbf{p} - \mathbf{q}|\sigma(\mathbf{q})\,dq;$$

$$\mathbf{p} \in B + \partial B_0 + \partial B_1,$$

$$\psi(\mathbf{p}) = \int_{\partial B_0} \theta(\mathbf{p} - \mathbf{q})\sigma(\mathbf{q})\,dq + \int_{\partial B_1} \theta(\mathbf{p} - \mathbf{q})\sigma(\mathbf{q})\,dq;$$

$$\mathbf{p} \in B + \partial B_0 + \partial B_1,$$

subject to the side condition

$$\int_{\partial B_1} \sigma(\mathbf{q})\,dq = 0 \qquad\qquad (4.5.18)$$

to ensure that ψ is single-valued in B. The angular potential

$$\int_{\partial B_0} \theta(\mathbf{p} - \mathbf{q})\,\sigma(\mathbf{q})\,dq$$

is the conjugate harmonic to $\int_{\partial B_0} \log|\mathbf{p} - \mathbf{q}|\sigma(\mathbf{q})\,dq$ within the entire interior domain enclosed by ∂B_0. Since this domain is simply-connected this harmonic function is necessarily single-valued. Substituting for ϕ into (4.5.17) yields a pair of coupled integral equations for σ, with the unknown constant ρ balanced by (4.5.18).

5

Vector Potential Theory

5.1 Introduction

Classical linear elastostatics may be formulated by a vector potential theory which closely parallels scalar potential theory. Thus the elastostatic displacement vector parallels the scalar harmonic function. The traction vector parallels the normal derivative. Well known global properties of the traction parallel the Gauss flux theorem. Vector potentials may be constructed which closely parallel the scalar simple- and double-layer potentials. Somigliana's formula parallels Green's formula. Vector boundary integral equations parallel the scalar boundary integral equations.

A direct symbolism, on the lines of dyadic symbolism (Weatherburn, 1962), is essential for exposing the analogies between scalar and vector formulae. However these analogies must be supported by detailed arguments, which are preferably presented through component equations utilising tensor symbolism and summation conventions. Our equations therefore often appear in these two complementary forms, which together bring out the theory to best advantage.

An elastostatic displacement vector is a continuous, at least twice-differentiable, vector

$$\boldsymbol{\phi}(\mathbf{p}) = \phi_\alpha(\mathbf{p}); \qquad \alpha = 1, 2, 3 \tag{5.1.1}$$

which defines the displacement field over some domain B of the elastic continuum. It is accompanied by a strain tensor

$$\tfrac{1}{2}(\phi_{\alpha/\beta} + \phi_{\beta/\alpha}); \qquad \alpha, \beta = 1, 2, 3 \tag{5.1.2}$$

64

where the solidus denotes differentiation with respect to the appropriate component of **p**, and by a stress tensor with components denoted $\Phi_{\alpha\beta}$. In a homogeneous linear elastic continuum, stress is related to strain by the generalised Hooke's law

$$\Phi_{\alpha\beta} = \tfrac{1}{2}C^{\alpha\beta\gamma\delta}(\phi_{\gamma/\delta} + \phi_{\delta/\gamma}) = C^{\alpha\beta\gamma\delta}\,\phi_{\gamma/\delta}; \qquad \alpha, \beta = 1, 2, 3 \qquad (5.1.3)$$

where γ, δ are dummy subscripts running through the values 1, 2, 3 and $C^{\alpha\beta\gamma\delta}$ are components of the elastic constant tensor characteristic of the medium. This is a fourth-rank tensor with $3^4 - 81$ components, which satisfy the relations

$$C^{\alpha\beta\gamma\delta} = C^{\gamma\delta\alpha\beta} = C^{\beta\alpha\gamma\delta}. \qquad (5.1.4)$$

By virtue of (5.1.4), the number of independent components becomes 21 for a medium having no special symmetry properties; this number reduces to 3 for a medium having cubic symmetry and to 2 for an isotropic medium.

Under the action of a body-force field $\mathbf{f}(\mathbf{p}) = f_\alpha(\mathbf{p})$; $\alpha = 1, 2, 3$ per unit volume, $\Phi_{\alpha\beta}$ satisfies Cauchy's equilibrium equations

$$\Phi_{\alpha\beta/\beta} + f_\alpha = 0; \qquad \alpha = 1, 2, 3, \qquad \mathbf{p} \in B. \qquad (5.1.5)$$

Substituting from (5.1.3) into (5.1.5) yields the field equations

$$C^{\alpha\beta\gamma\delta}\,\phi_{\gamma/\delta\beta} + f_\alpha = 0; \qquad \alpha = 1, 2, 3, \qquad \mathbf{p} \in B,$$

governing ϕ, which are preferably displayed as

$$C^{\gamma\delta\beta\alpha}\,\phi_{\gamma/\delta\beta} + f_\alpha = 0; \qquad \alpha = 1, 2, 3, \qquad \mathbf{p} \in B. \qquad (5.1.6)$$

These equations form an elliptic system (Birkhoff, 1971) which can only be solved within B subject to well posed conditions on the closed surface ∂B which bounds B. If ∂B is a Liapunov surface (see Appendix 1), there exists a traction vector $\phi^* = \phi_\alpha^*$; $\alpha = 1, 2, 3$ over ∂B, defined by

$$\phi_\alpha^*(\mathbf{p}) = \Phi_{\alpha\beta}(\mathbf{p})\,n_\beta(\mathbf{p}); \qquad \left.\begin{array}{l} \mathbf{p} \in \partial B \\ \alpha = 1, 2, 3 \end{array}\right\} \qquad (5.1.7)$$

at a point $\mathbf{p} \in \partial B$ where the inward unit normal vector has components

n_β; $\beta = 1, 2, 3$. This convention for the normal brings the vector formalism completely into line with the scalar formalism, but gives the traction vector an opposite sense to that usually adopted in elasticity theory. Two important integral relations satisfied by ϕ^*, which are essentially integrated versions of (5.1.5), are

$$\int_{\partial B} \phi^*(\mathbf{p}) . \mathbf{a} \, dp - \int_B \mathbf{f}(\mathbf{p}) . \mathbf{a} \, dP = 0, \tag{5.1.8}$$

$$\int_{\partial B} \mathbf{p} \wedge \phi^*(\mathbf{p}) . \mathbf{b} \, dp - \int_B \mathbf{p} \wedge \mathbf{f}(\mathbf{p}) . \mathbf{b} \, dP = 0, \tag{5.1.9}$$

where $\mathbf{a}$, $\mathbf{b}$ are arbitrary constant vectors. Relation (5.1.8) signifies that the resultant force produced by ϕ^* over ∂B balances the resultant force produced by $\mathbf{f}$ in B. Relation (5.1.9) signifies a corresponding theorem for the resultant moment. We prove (5.1.8) by noting that

$$\int_{\partial B} a_\alpha \phi_\alpha^* \, dp = \int_{\partial B} a_\alpha \Phi_{\alpha\beta} n_\beta \, dp = - \int_B a_\alpha \Phi_{\alpha\beta/\beta} \, dP = \int_B a_\alpha f_\alpha \, dP,$$

using the divergence theorem and (5.1.5). We prove (5.1.9) by writing $\mathbf{p} \wedge \phi^* . \mathbf{b} = \phi^* . \mathbf{b} \wedge \mathbf{p}$ and noting that

$$\int_{\partial B} (\mathbf{b} \wedge \mathbf{p})_\alpha \phi_\alpha^* \, dp = \int_{\partial B} (\mathbf{b} \wedge \mathbf{p})_\alpha \Phi_{\alpha\beta} n_\beta \, dp$$

$$= - \int_B \{ (\mathbf{b} \wedge \mathbf{p})_\alpha \Phi_{\alpha\beta/\beta} + (\mathbf{b} \wedge \mathbf{p})_{\alpha/\beta} \Phi_{\alpha\beta} \} \, dP$$

$$= \int_B (\mathbf{b} \wedge \mathbf{p})_\alpha f_\alpha \, dP,$$

bearing in mind

$$\Phi_{\alpha\beta} = \Phi_{\beta\alpha}, \qquad (\mathbf{b} \wedge \mathbf{p})_{\alpha/\beta} + (\mathbf{b} \wedge \mathbf{p})_{\beta/\alpha} = 0.$$

If $\mathbf{f} = 0$ everywhere in B, we refer to ϕ as a source-free field in B. This satisfies the homogeneous system of equations

$$C^{\gamma\delta\beta\alpha} \phi_{\gamma/\delta\beta} = 0; \qquad \alpha = 1, 2, 3. \tag{5.1.10}$$

Also, the accompanying tractions satisfy

$$\int_{\partial B} \boldsymbol{\phi}^*(\mathbf{p}) \cdot \mathbf{a}\, dp = 0, \qquad \int_{\partial B} \mathbf{p} \wedge \boldsymbol{\phi}^*(\mathbf{p}) \cdot \mathbf{b}\, dp = 0, \qquad (5.1.11)$$

showing that they constitute a self-equilibrated system of forces over ∂B. Conversely, if $\boldsymbol{\phi}^*$ has arbitrary prescribed values over ∂B, these must satisfy the integral relations (5.1.11) to be derivable from a source-free field in B.

5.2 Fundamental displacement dyadic

The most important example of a body-force field is a concentrated force acting at a specific point of the medium in a specific direction. A body force of magnitude 4π acting in the η-direction at $\mathbf{q}$ may be formally represented by the field vector with components

$$f_\alpha(\mathbf{p}) = 4\pi\delta(\mathbf{p}_\alpha, \mathbf{q}_\eta); \qquad \left.\begin{array}{l} \alpha = 1, 2, 3 \\ \eta = 1 \text{ or } 2 \text{ or } 3 \end{array}\right\}, \qquad (5.2.1)$$

where $\delta(\mathbf{p}_\alpha, \mathbf{q}_\eta)$ is a Dirac delta function sufficiently defined by the properties

$$\left.\begin{array}{c} \delta(\mathbf{p}_\alpha, \mathbf{q}_\eta) = 0 \text{ if } \mathbf{p} \neq \mathbf{q} \text{ or } \alpha \neq \eta \\[2mm] \delta(\mathbf{p}_\alpha, \mathbf{q}_\eta) \to \infty \text{ as } \mathbf{p} \to \mathbf{q}, \alpha - \eta \\[2mm] \left.\begin{array}{c} \displaystyle\int_B \delta(\mathbf{p}_\alpha, \mathbf{q}_\eta)dP = 1 \text{ if } \alpha = \eta \\[2mm] = 0 \text{ if } \alpha \neq \eta \end{array}\right\}, \qquad \mathbf{q} \in B \end{array}\right\}. \qquad (5.2.2)$$

This generates a displacement vector with components

$$\phi_\alpha(\mathbf{p}) = g(\mathbf{p}_\alpha, \mathbf{q}_\eta); \qquad \alpha = 1, 2, 3 \qquad (5.2.3)$$

which becomes singular in the component $\alpha = \eta$ when the field point $\mathbf{p}$ coincides with the source point $\mathbf{q}$. It may be noted that the subscripts α, η are tensor double subscripts which have been brought inside the brackets to indicate their association with $\mathbf{p}, \mathbf{q}$ respectively. This symbolism is preferable for our purposes to the alternative symbolism $g(\mathbf{p}, \mathbf{q})_{\alpha\eta}$.

Following (5.1.3) we write

$$G(\mathbf{p}_{\alpha\beta}, \mathbf{q}_\eta) = C^{\alpha\beta\gamma\delta} g(\mathbf{p}_{\gamma/\delta}, \mathbf{q}_\eta); \qquad \alpha, \beta = 1, 2, 3 \tag{5.2.4}$$

for the stress tensor accompanying (5.2.3), and equations (5.1.5) now read

$$G(\mathbf{p}_{\alpha\beta/\beta}, \mathbf{q}_\eta) + 4\pi\delta(\mathbf{p}_\alpha, \mathbf{q}_\eta) = 0; \qquad \alpha = 1, 2, 3. \tag{5.2.5}$$

Substituting from (5.2.4) into (5.2.5) yields the field equations

$$C^{\gamma\delta\beta\alpha} g(\mathbf{p}_{\gamma/\delta\beta}, \mathbf{q}_\eta) + 4\pi\delta(\mathbf{p}_\alpha, \mathbf{q}_\eta) = 0; \qquad \alpha = 1, 2, 3 \tag{5.2.6}$$

governing $g(\mathbf{p}_\alpha, \mathbf{q}_\eta)$, for which explicit solutions have been found in the isotropic case (Chapter 8) and in the plane anisotropic case (Kupradze, 1965). (The solution of (5.2.6) in the general anisotropic case has been discussed by Synge (1957).) Equations (5.2.6) formally hold even at $\mathbf{p} = \mathbf{q}$, and it is therefore mathematically convenient to regard (5.2.3) as an elastostatic displacement vector even at $\mathbf{p} = \mathbf{q}$. Essentially this means that singular displacement vectors within B satisfy the same integral relations over ∂B as those which are continuous and differentiable everywhere within B. A rigorous study of this, from the point of view of generalised functions, has been made by Furuhashi (1973).

So far η has been held at some fixed value defining the direction of the concentrated force at $\mathbf{q}$. As η runs through 1, 2, 3 we obtain three such forces at $\mathbf{q}$, each generating a singular vector of the type (5.2.3). These vectors may be displayed as columns of the array

$$\mathbf{g}(\mathbf{p}, \mathbf{q}) = \begin{bmatrix} g(\mathbf{p}_1, \mathbf{q}_1) \, g(\mathbf{p}_1, \mathbf{q}_2) \, g(\mathbf{p}_1, \mathbf{q}_3) \\ g(\mathbf{p}_2, \mathbf{q}_1) \, g(\mathbf{p}_2, \mathbf{q}_2) \, g(\mathbf{p}_2, \mathbf{q}_3) \\ g(\mathbf{p}_3, \mathbf{q}_1) \, g(\mathbf{p}_3, \mathbf{q}_2) \, g(\mathbf{p}_3, \mathbf{q}_3) \end{bmatrix} \tag{5.2.7}$$

which constitutes the fundamental displacement dyadic of the medium. Our symbolism indicates that $\mathbf{g}(\mathbf{p}, \mathbf{q})$ plays an analogous role to the unit simple source potential $g(\mathbf{p}, \mathbf{q})$ of the scalar theory. In line with $g(\mathbf{p}, \mathbf{q}) = g(\mathbf{q}, \mathbf{p})$ we shall prove (Chapter 6) that

$$g(\mathbf{p}_\alpha, \mathbf{q}_\eta) = g(\mathbf{q}_\eta, \mathbf{p}_\alpha), \tag{5.2.8}$$

i.e. $g(\mathbf{p}_\alpha, \mathbf{q}_\eta)$ may be interpreted as the displacement in the η-direction at $\mathbf{q}$ generated by a concentrated force acting in the α-direction at $\mathbf{p}$. An immediate corollary is that each row of $\mathbf{g(p, q)}$ defines a singular displacement vector generated at $\mathbf{p}$. i.e. the vector

$$\phi_\eta(\mathbf{q}) = g(\mathbf{p}_\alpha, \mathbf{q}_\eta); \qquad \eta = 1, 2, 3 \qquad (5.2.9)$$

with an accompanying stress tensor

$$G(\mathbf{p}_\alpha, \mathbf{q}_\eta) = g(\mathbf{p}_\alpha, \mathbf{q}_{\gamma/\delta})C^{\gamma\delta\eta\varepsilon}; \qquad \eta, \varepsilon = 1, 2, 3. \qquad (5.2.10)$$

5.3 Fundamental traction dyadics

Following (5.1.7) we write

$$g^*(\mathbf{p}_\alpha, \mathbf{q}_\eta) = G(\mathbf{p}_{\alpha\beta}, \mathbf{q}_\eta)n_\beta(\mathbf{p}); \qquad \alpha = 1, 2, 3 \qquad (5.3.1)$$

for the traction vector at $\mathbf{p} \in \partial B$ generated by a concentrated force acting in the η-direction at $\mathbf{q} \in B$. As η runs through 1, 2, 3 we obtain three such traction vectors at $\mathbf{p}$, which may be displayed as columns of the array

$$\mathbf{g^*(p, q)} = \begin{bmatrix} g^*(\mathbf{p}_1, \mathbf{q}_1)\, g^*(\mathbf{p}_1, \mathbf{q}_2)\, g^*(\mathbf{p}_1, \mathbf{q}_3) \\ g^*(\mathbf{p}_2, \mathbf{q}_1)\, g^*(\mathbf{p}_2, \mathbf{q}_2)\, g^*(\mathbf{p}_2, \mathbf{q}_3) \\ g^*(\mathbf{p}_3, \mathbf{q}_1)\, g^*(\mathbf{p}_3, \mathbf{q}_2)\, g^*(\mathbf{p}_3, \mathbf{q}_3) \end{bmatrix}. \qquad (5.3.2)$$

This is the fundamental traction dyadic of the medium, which plays an analogous role to the scalar normal derivative $g'(\mathbf{p}, \mathbf{q})$. Similarly, the traction dyadic

$$\mathbf{g(p, q)^*} = \begin{bmatrix} g(\mathbf{p}_1, \mathbf{q}_1)^*\ g(\mathbf{p}_1, \mathbf{q}_2)^*\ g(\mathbf{p}_1, \mathbf{q}_3)^* \\ g(\mathbf{p}_2, \mathbf{q}_1)^*\ g(\mathbf{p}_2, \mathbf{q}_2)^*\ g(\mathbf{p}_2, \mathbf{q}_3)^* \\ g(\mathbf{p}_3, \mathbf{q}_1)^*\ g(\mathbf{p}_3, \mathbf{q}_2)^*\ g(\mathbf{p}_3, \mathbf{q}_3)^* \end{bmatrix} \qquad (5.3.3)$$

parallels $g(\mathbf{p}, \mathbf{q})'$. An important property of $\mathbf{g(p, q)^*}$ is that each column defines a singular elastostatic displacement vector, i.e. the vector

$$g(\mathbf{p}_\alpha, \mathbf{q}_\eta)^*; \qquad \alpha = 1, 2, 3 \tag{5.3.4}$$

generated by a unit traction source acting in the η-direction at $\mathbf{q}$. To prove this we note that the accompanying stress tensor is

$$G(\mathbf{p}_{\alpha\beta}, \mathbf{q}_\eta)^* = C^{\alpha\beta\gamma\delta} g(\mathbf{p}_{\gamma/\delta}, \mathbf{q}_\eta)^*; \qquad \alpha, \beta = 1, 2, 3. \tag{5.3.5}$$

which satisfies (see Appendix 6) Cauchy's equilibrium equations

$$\frac{\partial}{\partial p_\beta} [G(\mathbf{p}_{\alpha\beta}, \mathbf{q}_\eta)^*] = -4\pi\delta (\mathbf{p}_\alpha, \mathbf{q}_\eta)^* \tag{5.3.6}$$

$$= 0 \text{ if } \mathbf{p} \neq \mathbf{q} \text{ or } \alpha \neq \eta,$$

where

$$\delta(\mathbf{p}_\alpha, \mathbf{q}_\eta)^* = \delta(\mathbf{p}_\alpha, \mathbf{q}_{\gamma/\delta}) C^{\gamma\delta\eta\varepsilon} n_\varepsilon(\mathbf{q}). \tag{5.3.7}$$

Accordingly $\mathbf{g}(\mathbf{p}, \mathbf{q})^*$ may be regarded as the displacement dyadic generated by unit traction sources at $\mathbf{q}$. Within a suitable context, therefore, it carries a parallel significance to the unit double source potential of the scalar theory.

Clearly each row of $\mathbf{g}^*(\mathbf{p}, \mathbf{q})$ defines a singular elastostatic displacement vector, i.e. the vector

$$g^*(\mathbf{p}_\alpha, \mathbf{q}_\eta); \qquad \eta = 1, 2, 3 \tag{5.3.8}$$

generated by a unit traction source acting in the α-direction at $\mathbf{p}$. Reciprocally each row of $\mathbf{g}(\mathbf{p}, \mathbf{q})^*$ defines a traction vector at $\mathbf{q}$, i.e. the vector

$$g(\mathbf{p}_\alpha, \mathbf{q}_\eta)^*; \qquad \eta = 1, 2, 3 \tag{5.3.9}$$

generated by a concentrated force acting in the α-direction at $\mathbf{p}$. It will be noted that any individual component of the dyadics (5.3.2), (5.3.3) carries two possible interpretations, i.e. it is either a traction component generated by a point force or a displacement component generated by a traction source. However the interpretation is always clear from the context as will be seen below.

It is sometimes convenient to introduce the fundamental dyadics

$$\mathbf{g}(\mathbf{q}, \mathbf{p}) = \begin{bmatrix} g(\mathbf{q}_1, \mathbf{p}_1)\, g(\mathbf{q}_1, \mathbf{p}_2)\, g(\mathbf{q}_1, \mathbf{p}_3) \\ g(\mathbf{q}_2, \mathbf{p}_1)\, g(\mathbf{q}_2, \mathbf{p}_2)\, g(\mathbf{q}_2, \mathbf{p}_3) \\ g(\mathbf{q}_3, \mathbf{p}_1)\, g(\mathbf{q}_3, \mathbf{p}_2)\, g(\mathbf{q}_3, \mathbf{p}_3) \end{bmatrix} \tag{5.3.10}$$

$$\mathbf{g}^*(\mathbf{q}, \mathbf{p}) = \begin{bmatrix} g^*(\mathbf{q}_1, \mathbf{p}_1)\, g^*(\mathbf{q}_1, \mathbf{p}_2)\, g^*(\mathbf{q}_1, \mathbf{p}_3) \\ g^*(\mathbf{q}_2, \mathbf{p}_1)\, g^*(\mathbf{q}_2, \mathbf{p}_2)\, g^*(\mathbf{q}_2, \mathbf{p}_3) \\ g^*(\mathbf{q}_3, \mathbf{p}_1)\, g^*(\mathbf{q}_3, \mathbf{p}_2)\, g^*(\mathbf{q}_3, \mathbf{p}_3) \end{bmatrix} \tag{5.3.11}$$

$$\mathbf{g}(\mathbf{q}, \mathbf{p})^* = \begin{bmatrix} g(\mathbf{q}_1, \mathbf{p}_1)^*\, g(\mathbf{q}_1, \mathbf{p}_2)^*\, g(\mathbf{q}_1, \mathbf{p}_3)^* \\ g(\mathbf{q}_2, \mathbf{p}_1)^*\, g(\mathbf{q}_2, \mathbf{p}_2)^*\, g(\mathbf{q}_2, \mathbf{p}_3)^* \\ g(\mathbf{q}_3, \mathbf{p}_1)^*\, g(\mathbf{q}_3, \mathbf{p}_2)^*\, g(\mathbf{q}_3, \mathbf{p}_3)^* \end{bmatrix} \tag{5.3.12}$$

These are seen to be the transpose dyadics to $\mathbf{g}(\mathbf{p}, \mathbf{q}), \mathbf{g}(\mathbf{p}, \mathbf{q})^*, \mathbf{g}^*(\mathbf{p}, \mathbf{q})$ respectively on bearing in mind (5.2.8).

5.4 Vector potentials

Corresponding to the scalar simple-source density $\sigma(\mathbf{q})$ of Chapter 1, we introduce the vector simple-source density

$$\sigma(\mathbf{q}) = \sigma_\eta(\mathbf{q}); \qquad \left.\begin{matrix} \mathbf{q} \in \partial B \\ \eta = 1, 2, 3 \end{matrix}\right\}. \tag{5.4.1}$$

This enables us to construct a vector simple-layer potential corresponding to (1.3.1), viz.

$$\mathbf{V}(\mathbf{p}) = \int_{\partial B} \mathbf{g}(\mathbf{p}, \mathbf{q}) \cdot \sigma(\mathbf{q})\, d\mathbf{q} = \int_{\partial B} \sigma(\mathbf{q}) \cdot \mathbf{g}(\mathbf{q}, \mathbf{p})\, d\mathbf{q} \tag{5.4.2}$$

with components

$$V_\alpha(\mathbf{p}) = \int_{\partial B} g(\mathbf{p}_\alpha, \mathbf{q}_\eta) \sigma_\eta(\mathbf{q}) \, dq = \int_{\partial B} \sigma_\eta(\mathbf{q}) g(\mathbf{q}_\eta, \mathbf{p}_\alpha) \, dq; \qquad \alpha = 1, 2, 3. \qquad (5.4.3)$$

If ∂B is a Liapunov surface (not necessarily closed), and if σ is Hölder-continuous over ∂B, then $\mathbf{V}$ is an elastostatic displacement vector everywhere except at ∂B. For an isotropic medium, and for a plane anisotropic medium, $\mathbf{V}$ remains continuous at ∂B (Kupradze, 1965). We shall build up a consistent theory for the general anisotropic medium on the assumption that $\mathbf{V}$ remains continuous in this case also.

If ∂B is a closed surface, we must distinguish between the interior potential generated by σ and the exterior potential generated by σ. By analogy with (1.3.7) and (1.3.8), the accompanying traction vectors appear as

$$V_i^*(\mathbf{p}) = \int_{\partial B} \mathbf{g}_i^*(\mathbf{p}, \mathbf{q}) \cdot \sigma(\mathbf{q}) \, dq - 2\pi\sigma(\mathbf{p}); \qquad \mathbf{p} \in \partial B, \qquad (5.4.4)$$

$$V_e^*(\mathbf{p}) = \int_{\partial B} \mathbf{g}_e^*(\mathbf{p}, \mathbf{q}) \cdot \sigma(\mathbf{q}) \, dq - 2\pi\sigma(\mathbf{p}); \qquad \mathbf{p} \in \partial B, \qquad (5.4.5)$$

where $\mathbf{g}_i^*$, $\mathbf{g}_e^*$ are given by (5.3.2) using the appropriate normals. These formulae may be proved rigorously for the isotropic and plane aniso-tropic media (Kupradze, 1965). Since

$$\mathbf{g}_i^*(\mathbf{p}, \mathbf{q}) + \mathbf{g}_e^*(\mathbf{p}, \mathbf{q}) = 0, \qquad (5.4.6)$$

it follows that

$$\mathbf{V}_i^* + \mathbf{V}_e^* = -4\pi\sigma. \qquad (5.4.7)$$

Corresponding to the scalar double-source density $\mu(\mathbf{q})$, we introduce the vector double-source density

$$\mu(\mathbf{q}) = \mu_\eta(\mathbf{q}); \qquad \left. \begin{array}{l} \mathbf{q} \in \partial B \\ \eta = 1, 2, 3 \end{array} \right\}. \qquad (5.4.8)$$

This enables us to construct a vector double-layer potential corresponding to (1.4.4), viz.

$$\mathbf{W}(\mathbf{p}) = \int_{\partial B} \mathbf{g}(\mathbf{p}, \mathbf{q})_i^* \cdot \boldsymbol{\mu}(\mathbf{q}) \, dq = \int_{\partial B} \boldsymbol{\mu}(\mathbf{q}) \cdot \mathbf{g}_i^*(\mathbf{q}, \mathbf{p}) \, dq, \qquad (5.4.9)$$

where $\mathbf{g}(\mathbf{p}, \mathbf{q})_i^*$ is given by (5.3.3) using the appropriate normal. If ∂B is a Liapunov surface (not necessarily closed), and if μ is continuous over ∂B, then $\mathbf{W}$ is an elastostatic displacement vector everywhere except at ∂B. Here it suffers jumps defined by

$$\lim_{\mathbf{p}_i \to \mathbf{p}} \mathbf{W}(\mathbf{p}_i) = \mathbf{W}(\mathbf{p}) + 2\pi\boldsymbol{\mu}(\mathbf{p}); \qquad \mathbf{p} \in \partial B, \qquad (5.4.10)$$

$$\lim_{\mathbf{p}_e \to \mathbf{p}} \mathbf{W}(\mathbf{p}_e) = \mathbf{W}(\mathbf{p}) - 2\pi\boldsymbol{\mu}(\mathbf{p}); \qquad \mathbf{p} \in \partial B. \qquad (5.4.11)$$

By analogy with the normal derivative properties (1.4.8) of the scalar theory, we may assume the continuity of the traction vector $\mathbf{W}^*$ associated with $\mathbf{W}$. Similarly, by analogy with the tangential derivative properties (1.4.9) of the scalar theory, we may assume corresponding jumps in the appropriate traction vectors. Similar properties hold for $\mathbf{W}$ corresponding to (1.4.13).

Two particular choices of μ are of interest when ∂B is closed, viz. $\mu = \mathbf{a}$ and $\mu = \mathbf{b} \wedge \mathbf{q}$ where $\mathbf{a}, \mathbf{b}$ are arbitrary constant vectors. These generate potentials

$$\mathbf{W}(\mathbf{p}) = \int_{\partial B} \mathbf{g}(\mathbf{p}, \mathbf{q})_i^* \cdot \mathbf{a} \, dq = 4\pi\mathbf{a}; \qquad \mathbf{p} \in B_i, \qquad (5.4.12)$$

$$\mathbf{W}(\mathbf{p}) = \int_{\partial B} \mathbf{g}(\mathbf{p}, \mathbf{q})_i^* \cdot \mathbf{b} \wedge \mathbf{q} \, dq = 4\pi\mathbf{b} \wedge \mathbf{p}; \qquad \mathbf{p} \in B_i, \qquad (5.4.13)$$

which results are essentially vector generalisations of the Gauss flux theorem. We prove (5.4.12) by introducing

$$\phi(\mathbf{p}) = g(\mathbf{p}_\alpha, \mathbf{q}_\eta), \qquad \mathbf{f}(\mathbf{p}) = 4\pi\delta(\mathbf{p}_\alpha, \mathbf{q}_\eta); \qquad \alpha = 1, 2, 3$$

$$\eta = 1 \text{ or } 2 \text{ or } 3$$

into (5.1.8), which gives (in component form)

$$\int_{\partial B} g_i^*(\mathbf{p}_\alpha, \mathbf{q}_\eta) a_\alpha \, dp = 4\pi \int_{B_i} \delta(\mathbf{p}_\alpha, \mathbf{q}_\eta) a_\alpha \, dP; \qquad \mathbf{p} \in B_i + \partial B$$

$$\mathbf{q} \in B_i$$

i.e. $\displaystyle\int_{\partial B} g(\mathbf{q}_\eta, \mathbf{p}_\alpha)_i^* a_\alpha \, dp = 4\pi a_\eta; \qquad \mathbf{q} \in B_i.$ \hfill (5.4.14)

Equation (5.4.14) holds for $\eta = 1, 2, 3$ and it therefore appears in vector form as

$$\int_{\partial B} \mathbf{g}(\mathbf{q}, \mathbf{p})_i^* \cdot \mathbf{a} \, dp = 4\pi\mathbf{a}; \qquad \mathbf{q} \in B_i, \tag{5.4.15}$$

which is an obvious adaptation of (5.4.12). A similar proof holds for (5.4.13). It follows from (5.4.12), by virtue of (5.4.10), (5.4.11) that

$$\int_{\partial B} \mathbf{g}(\mathbf{p}, \mathbf{q})_i^* \cdot \mathbf{a} \, dq = 2\pi\mathbf{a}; \qquad \mathbf{p} \in \partial B, \tag{5.4.16}$$

$$\int_{\partial B} \mathbf{g}(\mathbf{p}, \mathbf{q})_i^* \cdot \mathbf{a} \, dq = 0; \qquad \mathbf{p} \in B_e. \tag{5.4.17}$$

Similarly

$$\int_{\partial B} \mathbf{g}(\mathbf{p}, \mathbf{q})_i^* \cdot \mathbf{b} \wedge \mathbf{q} \, dq = 2\pi\mathbf{b} \wedge \mathbf{p}; \qquad \mathbf{p} \in \partial B, \tag{5.4.18}$$

$$\int_{\partial B} \mathbf{g}(\mathbf{p}, \mathbf{q})_i^* \cdot \mathbf{b} \wedge \mathbf{q} \, dq = 0; \qquad \mathbf{p} \in B_e. \tag{5.4.19}$$

6

Somigliana's Formula

6.1 Asymptotic behaviour of vector potentials

Formula (5.4.5) provides the exterior traction vector $\mathbf{V}_e^*(\mathbf{p})$ at $\mathbf{p} \in \partial B$ associated with the displacement vector

$$\mathbf{V}(\mathbf{p}) = \int_{\partial B} \boldsymbol{\sigma}(\mathbf{q}) \cdot \mathbf{g}(\mathbf{q}, \mathbf{p}) \, dq; \qquad \mathbf{p} \in B_e + \partial B.$$

It is of interest to examine the force and moment resultants produced by these tractions over ∂B. We find

$$\int_{\partial B} \mathbf{V}_e^*(\mathbf{p}) \cdot \mathbf{a} \, dp = \int_{\partial B} \left\{ \int_{\partial B} \boldsymbol{\sigma}(\mathbf{q}) \cdot \mathbf{g}(\mathbf{q}, \mathbf{p})_e^* \, dq \right\} \cdot \mathbf{a} \, dp - 2\pi \int_{\partial B} \boldsymbol{\sigma}(\mathbf{p}) \cdot \mathbf{a} \, dp$$

$$= \int_{\partial B} \boldsymbol{\sigma}(\mathbf{q}) \cdot \left\{ \int_{\partial B} \mathbf{g}(\mathbf{q}, \mathbf{p})_e^* \cdot \mathbf{a} \, dp \right\} dq - 2\pi \int_{\partial B} \boldsymbol{\sigma}(\mathbf{p}) \cdot \mathbf{a} \, dp$$

$$= - \int_{\partial B} \boldsymbol{\sigma}(\mathbf{q}) \cdot (2\pi \mathbf{a}) \, dq - 2\pi \int_{\partial B} \boldsymbol{\sigma}(\mathbf{p}) \cdot \mathbf{a} \, dp$$

$$= - 4\pi \int_{\partial B} \boldsymbol{\sigma}(\mathbf{q}) \cdot \mathbf{a} \, dq, \tag{6.1.1}$$

assuming that the order of integration may be interchanged and using a slight adaptation of (5.4.16). Similarly we find

$$\int_{\partial B} \mathbf{V}_e^*(\mathbf{p}) \cdot \mathbf{b} \wedge \mathbf{p} \, dp = -4\pi \int_{\partial B} \boldsymbol{\sigma}(\mathbf{q}) \cdot \mathbf{b} \wedge \mathbf{q} \, dq = -4\pi \int_{\partial B} \mathbf{q} \wedge \boldsymbol{\sigma}(\mathbf{q}) \cdot \mathbf{b} \, dq,$$

(6.1.2)

using a slight adaptation of (5.4.18). These resultants must be balanced by compensating resultants at infinity in order to ensure that B_e remains in statical equilibrium. To examine this, we enclose ∂B by a large spherical surface ∂S and note that

$$\mathbf{V}^*(\mathbf{p}) = \int_{\partial B} \boldsymbol{\sigma}(\mathbf{q}) \cdot \mathbf{g}(\mathbf{q}, \mathbf{p})^* \, dq; \qquad \mathbf{p} \in \partial S, \qquad (6.1.3)$$

where $\mathbf{g}(\mathbf{q}, \mathbf{p})^*$ signifies the (inward) traction dyadic at $\mathbf{p} \in \partial S$ generated by $\boldsymbol{\sigma}(\mathbf{q})$, $\mathbf{q} \in \partial B$. It follows without difficulty, using slight adaptations of (5.4.12), (5.4.13), that

$$\int_{\partial S} \mathbf{V}^*(\mathbf{p}) \cdot \mathbf{a} \, dp = 4\pi \int_{\partial B} \boldsymbol{\sigma}(\mathbf{q}) \cdot \mathbf{a} \, dq, \qquad (6.1.4)$$

$$\int_{\partial S} \mathbf{V}^*(\mathbf{p}) \cdot \mathbf{b} \wedge \mathbf{p} \, dp = 4\pi \int_{\partial B} \mathbf{q} \wedge \boldsymbol{\sigma}(\mathbf{q}) \cdot \mathbf{b} \, dq, \qquad (6.1.5)$$

as expected.

From the existence of the integral

$$\int_{\partial S} \mathbf{g}(\mathbf{q}, \mathbf{p})^* \cdot \mathbf{a} \, dp = 4\pi \mathbf{a}; \qquad \mathbf{q} \in \partial B, \qquad (6.1.6)$$

we may infer that $\mathbf{g}(\mathbf{q}, \mathbf{p})^* = O(|\mathbf{p}|^{-2})$ as $|\mathbf{p}| \to \infty$ for a fixed $\mathbf{q}$, whence $\mathbf{g}(\mathbf{q}, \mathbf{p}) = O(|\mathbf{p}|^{-1})$ as $|\mathbf{p}| \to \infty$. (An alternative argument leading to the same conclusion may be based upon the system of equations (5.2.6).) An application of Taylor's theorem then yields the expansion

$$\mathbf{g}(\mathbf{q}, \mathbf{p}) = \mathbf{g}(\mathbf{0}, \mathbf{p}) + \mathbf{q} \cdot \nabla \mathbf{g}(\mathbf{0}, \mathbf{p}) + O(|\mathbf{p}|^{-3}) \quad \text{as} \quad |\mathbf{p}| \to \infty, \quad (6.1.7)$$

where $\nabla \mathbf{g}(\mathbf{0}, \mathbf{p})$ denotes the gradient vector at $\mathbf{q} = 0$ associated with each component of $\mathbf{g(q, p)}$. Each term of this expansion makes its specific contribution to the force-resultant integral (6.1.6), thus

$$\int_{\partial S} \mathbf{g(0, p)}^* . \mathbf{a} \, dp = 4\pi\mathbf{a}, \tag{6.1.8}$$

$$\int_{\partial S} [\mathbf{q} . \nabla \mathbf{g(0, p)}^*] . \mathbf{a} \, dp = 4\pi\mathbf{q} . \nabla \mathbf{a} = 0. \tag{6.1.9}$$

Also they make their specific contributions to the moment-resultant integral

$$\int_{\partial S} \mathbf{g(q, p)}^* . \mathbf{b} \wedge \mathbf{p} \, dp = 4\pi\mathbf{b} \wedge \mathbf{q}, \qquad \mathbf{q} \in \partial B, \tag{6.1.10}$$

thus

$$\int_{\partial S} \mathbf{g(0, p)}^* . \mathbf{b} \wedge \mathbf{p} \, dp = 4\pi(\mathbf{b} \wedge \mathbf{q})_{\mathbf{q}=0} = 0, \tag{6.1.11}$$

$$\int_{\partial S} [\mathbf{q} . \nabla \mathbf{g(0, p)}^*] . \mathbf{b} \wedge \mathbf{p} \, dp = 4\pi\mathbf{q} . \nabla(\mathbf{b} \wedge \mathbf{q})_{\mathbf{q}=0} = 4\pi\mathbf{b} \wedge \mathbf{q}. \tag{6.1.12}$$

No contributions either to the force or moment resultants arise from the $O(|\mathbf{p}|^{-3})$ terms of (6.1.7). The results (6.1.9), (6.1.12) are proved directly in Appendix 6, but they are also seen to be true by subtracting, respectively, (6.1.8) from (6.1.6) and (6.1.11) from (6.1.10), bearing in mind (6.1.7).

Corresponding to the expansion (6.1.7), we introduce the expansions

$$\mathbf{V(p)} = \int_{\partial B} \sigma(\mathbf{q}) . \mathbf{g(0, p)} \, dq + \int_{\partial B} \sigma(\mathbf{q}) . [\mathbf{q} . \nabla \mathbf{g(0, p)}] \, dq + O(|\mathbf{p}|^{-3})$$
$$\text{as} \quad |\mathbf{p}| \to \infty, \tag{6.1.13}$$

$$\mathbf{V}^*(\mathbf{p}) = \int_{\partial B} \sigma(\mathbf{q}) . \mathbf{g(0, p)}^* \, dq + \int_{\partial B} \sigma(\mathbf{q}) . [\mathbf{q} . \nabla \mathbf{g(0, p)}^*] \, dq + O(|\mathbf{p}|^{-4})$$
$$\text{as} \quad |\mathbf{p}| \to \infty. \tag{6.1.14}$$

It is evident that only the first integral of (6.1.14) contributes to (6.1.4), and that only the second integral contributes to (6.1.5).

The fundamental traction dyadic $\mathbf{g}_e^*(\mathbf{q, p})$ may be interpreted as the displacement dyadic at $\mathbf{p}$ generated by unit traction sources at $\mathbf{q}$. This is accompanied by a traction vector $\mathbf{g}_e^*(\mathbf{q}, \mathbf{p})^*$ at $\mathbf{p} \in \partial S$, which produces the force and moment resultants

$$\int_{\partial S} \mathbf{g}_e^*(\mathbf{q}, \mathbf{p})^* \cdot \mathbf{a} \, dp = 0; \qquad \mathbf{q} \in \partial B, \tag{6.1.15}$$

$$\int_{\partial S} \mathbf{g}_e^*(\mathbf{q}, \mathbf{p})^* \cdot \mathbf{b} \wedge \mathbf{p} \, dp = 0; \qquad \mathbf{q} \in \partial B, \tag{6.1.16}$$

as follows from Appendix 6. More generally, the vector double-layer potential

$$\mathbf{W}(\mathbf{p}) = \int_{\partial B} \boldsymbol{\mu}(\mathbf{q}) \cdot \mathbf{g}_e^*(\mathbf{q}, \mathbf{p}) \, dq \tag{6.1.17}$$

is accompanied by a traction vector

$$\mathbf{W}^*(\mathbf{p}) = \int_{\partial B} \boldsymbol{\mu}(\mathbf{q}) \cdot \mathbf{g}_e^*(\mathbf{q}, \mathbf{p})^* \, dq; \qquad \mathbf{p} \in \partial S, \tag{6.1.18}$$

which produces the force and moment resultants

$$\int_{\partial S} \mathbf{W}^*(\mathbf{p}) \cdot \mathbf{a} \, dp = \int_{\partial B} \boldsymbol{\mu}(\mathbf{q}) \cdot \left\{ \int_{\partial S} \mathbf{g}_e^*(\mathbf{q}, \mathbf{p})^* \cdot \mathbf{a} \, dp \right\} dq = 0, \tag{6.1.19}$$

$$\int_{\partial S} \mathbf{W}^*(\mathbf{p}) \cdot \mathbf{b} \wedge \mathbf{p} \, dp = \int_{\partial B} \boldsymbol{\mu}(\mathbf{q}) \cdot \left\{ \int_{\partial S} \mathbf{g}_e^*(\mathbf{q}, \mathbf{p})^* \cdot \mathbf{b} \wedge \mathbf{p} \, dp \right\} dq = 0. \tag{6.1.20}$$

These results show that $\mathbf{W}$, by contrast with $\mathbf{V}$, cannot represent a ϕ in B_e characterised by a non-zero force or moment resultant over ∂B. An analogous limitation on the scalar double-layer potential has been noted in Section 2.6. Nothing is gained by introducing the asymptotic expansion

$$\mathbf{g}_e^*(\mathbf{q}, \mathbf{p}) = \mathbf{g}_e^*(\mathbf{0}, \mathbf{p}) + O(|\mathbf{p}|^{-3}) \quad \text{as} \quad |\mathbf{p}| \to \infty, \tag{6.1.21}$$

because the asymptotic properties of $\mathbf{g}_e^*(\mathbf{0}, \mathbf{p})$ are essentially those of $\mathbf{g}_e^*(\mathbf{q}, \mathbf{p})$.

6.2 Betti-Somigliana formula: component form

Two elastostatic displacement fields

$$\phi_\eta, \psi_\eta; \qquad \eta = 1, 2, 3 \tag{6.2.1}$$

in $B_i + \partial B$, accompanied by stress fields

$$\Phi_{\eta\varepsilon}, \Psi_{\eta\varepsilon}; \qquad \eta, \varepsilon = 1, 2, 3 \tag{6.2.2}$$

in B_i, and by (inward) traction vectors

$$\phi_\eta^*, \psi_\eta^*; \qquad \eta = 1, 2, 3 \tag{6.2.3}$$

over ∂B, satisfy the general reciprocal relation

$$\int_{\partial B} [\phi_\eta \psi_\eta^* - \psi_\eta \phi_\eta^*] \, dq = - \int_{B_i} [\phi_\eta \Psi_{\eta\varepsilon/\varepsilon} - \psi_\eta \Phi_{\eta\varepsilon/\varepsilon}] \, dQ \tag{6.2.4}$$

which parallels the scalar relation (3.1.4). Thus,

$$\int_{\partial B} \phi_\eta \psi_\eta^* \, dq = \int_{\partial B} \phi_\eta \Psi_{\eta\varepsilon} n_\varepsilon \, dq = - \int_{B_i} \lceil \phi_\eta \Psi_{\eta\varepsilon/\varepsilon} + \phi_{\eta/\varepsilon} \Psi_{\eta\varepsilon} \rceil \, dQ, \tag{6.2.5}$$

$$\int_{\partial B} \psi_\eta \phi_\eta^* \, dq = \int_{\partial B} \psi_\eta \Phi_{\eta\varepsilon} n_\varepsilon \, dq = - \int_{B_i} [\psi_\eta \Phi_{\eta\varepsilon/\varepsilon} + \psi_{\eta/\varepsilon} \Phi_{\eta\varepsilon}] \, dQ, \tag{6.2.6}$$

by an application of the divergence theorem, where $n_\varepsilon; \varepsilon = 1, 2, 3$ is the unit inward normal vector at $\mathbf{q} \in \partial B$, whence (6.2.4) follows by subtraction bearing in mind

$$\int_{B_i} \phi_{\eta/\varepsilon} \Psi_{\eta\varepsilon} \, dQ = \int_{B_i} \psi_{\eta/\varepsilon} \Phi_{\eta\varepsilon} \, dQ. \tag{6.2.7}$$

If the displacement fields are source-free in B_i, then

$$\Phi_{\eta\varepsilon/\varepsilon} = 0, \qquad \Psi_{\eta\varepsilon/\varepsilon} = 0; \qquad \eta = 1, 2, 3$$

and the right-hand side of (6.2.4) vanishes, so yielding Betti's reciprocal relation

$$\int_{\partial B} [\phi_\eta \psi_\eta^* - \psi_\eta \phi_\eta^*] \, dq = 0. \tag{6.2.8}$$

If

$$\psi_\eta(\mathbf{q}) = g(\mathbf{q}_\eta, \mathbf{p}_\alpha); \qquad \left. \begin{array}{l} \mathbf{q} \in B_i + \partial B, \ \mathbf{p} \in B_i \\[1mm] \eta = 1, 2, 3 \end{array} \right\}, \tag{6.2.9}$$

whilst $\phi_\eta(\mathbf{q})$; $\eta = 1, 2, 3$ remains source-free in $B_i + \partial B$, then

$$\Phi_{\eta\varepsilon/\varepsilon}(\mathbf{q}) = 0, \qquad \Psi_{\eta\varepsilon/\varepsilon}(\mathbf{q}) = -4\pi\delta(\mathbf{q}_\eta, \mathbf{p}_\alpha); \qquad \eta = 1, 2, 3.$$

The divergence theorem formally applies in this case despite the singularity at $\mathbf{p}$. Accordingly, substituting into the right-hand side of (6.2.4), we obtain

$$4\pi \int_{B_i} \phi_\eta(\mathbf{q}) \, \delta(\mathbf{q}_\eta, \mathbf{p}_\alpha) \, dQ = 4\pi\phi_\alpha(\mathbf{p}); \qquad \mathbf{p} \in B_i, \tag{6.2.10}$$

from which follows Somigliana's formula

$$\int_{\partial B} [\phi_\eta(\mathbf{q}) \, g_i^*(\mathbf{q}_\eta, \mathbf{p}_\alpha) - g(\mathbf{q}_\eta, \mathbf{p}_\alpha) \, \phi_\eta^*(\mathbf{q})] \, dq = 4\pi\phi_\alpha(\mathbf{p}); \qquad \left. \begin{array}{l} \mathbf{p} \in B_i \\[1mm] \alpha = 1, 2, 3 \end{array} \right\}. \tag{6.2.11}$$

It may be remarked that the roles of $\mathbf{p}$ and $\mathbf{q}$ are interchanged in the passage from (6.2.9) to (6.2.11), i.e. $\mathbf{q}$ is the field variable in (6.2.9) but emerges as the source variable in (6.2.11) and reciprocally for $\mathbf{p}$. This is because the integrations with respect to $\mathbf{q}$ convert it from one role to the other, so automatically resulting in a reciprocal change for $\mathbf{p}$.

We now choose

$$\phi_\eta(\mathbf{q}) = g(\mathbf{q}_\eta, \mathbf{x}_\alpha), \qquad \psi_\eta(\mathbf{q}) = g(\mathbf{q}_\eta, \mathbf{y}_\beta); \qquad \left. \begin{array}{l} \mathbf{x}, \mathbf{y} \in B_i \\[1mm] \eta = 1, 2, 3 \end{array} \right\}, \qquad \mathbf{q} \in B_i + \partial B$$

$$\tag{6.2.12}$$

in which case

$$\Phi_{\eta\varepsilon/\varepsilon}(\mathbf{q}) = -4\pi\delta(\mathbf{q}_\eta, \mathbf{x}_\alpha), \qquad \Psi_{\eta\varepsilon/\varepsilon}(\mathbf{q}) = -4\pi\delta(\mathbf{q}_\eta, \mathbf{y}_\beta); \qquad \eta = 1, 2, 3.$$

The right-hand side of (6.2.4) accordingly becomes

$$4\pi \int_{B_i} [g(\mathbf{q}_\eta, \mathbf{x}_\alpha)\,\delta(\mathbf{q}_\eta, \mathbf{y}_\beta) - g(\mathbf{q}_\eta, \mathbf{y}_\beta)\,\delta(\mathbf{q}_\eta, \mathbf{x}_\alpha)]\,\mathrm{d}q$$

$$= 4\pi[g(\mathbf{y}_\beta, \mathbf{x}_\alpha) - g(\mathbf{x}_\alpha, \mathbf{y}_\beta)].$$

Also, the left-hand side of (6.2.4) now vanishes by an application (see footnote on p. 19 Chapter 1) of Betti's reciprocal relation (6.2.8) to B_e, regarded as bounded internally by ∂B and externally by a large spherical surface ∂S. Accordingly $g(\mathbf{y}_\beta, \mathbf{x}_\alpha) = g(\mathbf{x}_\alpha, \mathbf{y}_\beta)$ as assumed in (5.2.8).

Finally, putting $\phi_\eta = \psi_\eta$ in (6.2.5) or (6.2.6), we obtain the vector analogue of (3.1.11), viz.

$$\int_{\partial B} \phi_\eta \phi_\eta^* \, \mathrm{d}q = -\int_{B_i} \phi_{\eta/\varepsilon} \Phi_{\eta\varepsilon} \, \mathrm{d}Q, \tag{6.2.13}$$

which appears preferably in the form

$$-\int_{\partial B} \boldsymbol{\phi} \cdot \boldsymbol{\phi}^* \, \mathrm{d}q = \int_{B_i} \tfrac{1}{2}(\phi_{\eta/\varepsilon} + \phi_{\varepsilon/\eta}) \Phi_{\eta\varepsilon} \, \mathrm{d}Q. \tag{6.2.14}$$

The left-hand side of (6.2.14) defines the work done by the surface tractions in producing the surface displacements. The right-hand integrand defines the strain-energy density of the medium, and this is an inherently positive quantity for any stable elastic material (see Appendix 7).

6.3 Betti-Somigliana formula: vector form

Recalling that $g(\mathbf{q}_\eta, \mathbf{p}_\alpha) = g(\mathbf{p}_\alpha, \mathbf{q}_\eta)$, we may write Somigliana's formula (6.2.11) in the vector-dyadic form

$$\int_{\partial B} [\mathbf{g}(\mathbf{p}, \mathbf{q})_i^* \cdot \boldsymbol{\phi}(\mathbf{q}) - \mathbf{g}(\mathbf{p}, \mathbf{q}) \cdot \boldsymbol{\phi}_i^*(\mathbf{q})]\,\mathrm{d}q = 4\pi\boldsymbol{\phi}(\mathbf{p}); \qquad \mathbf{p} \in B_i. \tag{6.3.1}$$

This exhibits $4\pi\phi(\mathbf{p})$ in B_i as the superposition of a vector simple-layer potential

$$\int_{\partial B} \mathbf{g}(\mathbf{p}, \mathbf{q}) \cdot \phi_i^*(\mathbf{q}) \, dq; \qquad \mathbf{p} \in B_i, \tag{6.3.2}$$

and a vector double-layer potential

$$\int_{\partial B} \mathbf{g}(\mathbf{p}, \mathbf{q})_i^* \cdot \phi(\mathbf{q}) \, dq; \qquad \mathbf{p} \in B_i, \tag{6.3.3}$$

defined respectively by the vector source densities

$$\sigma = \phi_i^*, \qquad \mu = \phi \quad \text{over} \quad \partial B. \tag{6.3.4}$$

Clearly Somigliana's formula constitutes the complete vector analogue of Green's formula, with the traction vector playing the role of the scalar normal derivative.

The potential (6.3.2) remains continuous as $\mathbf{p}$ passes from B_i onto ∂B, but the potential (6.3.3) jumps by an amount $-2\pi\phi(\mathbf{p})$. Therefore the right-hand side of formula (6.3.1) becomes $2\pi\phi(\mathbf{p})$ on ∂B, so yielding Somigliana's boundary formula

$$\int_{\partial B} [\mathbf{g}(\mathbf{p}, \mathbf{q})_i^* \cdot \phi(\mathbf{q}) - \mathbf{g}(\mathbf{p}, \mathbf{q}) \cdot \phi_i^*(\mathbf{q})] \, dq = 2\pi\phi(\mathbf{p}); \qquad \mathbf{p} \in \partial B. \tag{6.3.5}$$

An alternative approach is to note that (6.2.10) gets replaced by

$$4\pi \int_{B_i} \phi_\eta(\mathbf{q}) \, \delta(\mathbf{q}_\eta, \mathbf{p}_\alpha) \, dQ = 2\pi\phi_\alpha(\mathbf{p}); \qquad \mathbf{p} \in \partial B \tag{6.3.6}$$

in this case, since $\mathbf{q} = \mathbf{p}$ at the boundary of the domain of integration. As $\mathbf{p}$ passes from ∂B into B_e, (6.2.10) gets replaced by

$$4\pi \int_{B_i} \phi_\eta(\mathbf{q}) \, \delta(\mathbf{q}_\eta, \mathbf{p}_\alpha) \, dQ = 0; \qquad \mathbf{p} \in B_e, \tag{6.3.7}$$

so yielding the interior reciprocal relation

$$\int_{\partial B} [\mathbf{g}(\mathbf{p}, \mathbf{q})_i^* \cdot \boldsymbol{\phi}(\mathbf{q}) - \mathbf{g}(\mathbf{p}, \mathbf{q}) \cdot \boldsymbol{\phi}_i^*(\mathbf{q})] \, dq = 0; \qquad \mathbf{p} \in B_e. \tag{6.3.8}$$

This may be viewed as a particular case of Betti's reciprocal relation (6.2.8), since both $\boldsymbol{\phi}_\eta(\mathbf{q})$, $g(\mathbf{q}_\eta, \mathbf{p}_\alpha)$ are source-free fields in $B_i + \partial B$ if $\mathbf{p} \in B_e$.

Changing i into e yields Somigliana's exterior formula

$$\int_{\partial B} [\mathbf{g}(\mathbf{p}, \mathbf{q})_e^* \cdot \boldsymbol{\phi}(\mathbf{q}) - \mathbf{g}(\mathbf{p}, \mathbf{q}) \cdot \boldsymbol{\phi}_e^*(\mathbf{q})] \, dq = 4\pi\boldsymbol{\phi}(\mathbf{p}); \qquad \mathbf{p} \in B_e, \tag{6.3.9}$$

which clearly only holds if $\boldsymbol{\phi} = O(|\mathbf{p}|^{-1})$ as $|\mathbf{p}| \to \infty$, i.e. $\boldsymbol{\phi}$ is regular at infinity. As $\mathbf{p}$ passes from B_e onto ∂B, we obtain the exterior boundary formula

$$\int_{\partial B} [\mathbf{g}(\mathbf{p}, \mathbf{q})_e^* \cdot \boldsymbol{\phi}(\mathbf{q}) - \mathbf{g}(\mathbf{p}, \mathbf{q}) \cdot \boldsymbol{\phi}_e^*(\mathbf{q})] \, dq = 2\pi\boldsymbol{\phi}(\mathbf{p}); \qquad \mathbf{p} \in \partial B. \tag{6.3.10}$$

Finally, as $\mathbf{p}$ passes from ∂B into B_i, there follows the exterior reciprocal relation

$$\int_{\partial B} [\mathbf{g}(\mathbf{p}, \mathbf{q})_e^* \cdot \boldsymbol{\phi}(\mathbf{q}) - \mathbf{g}(\mathbf{p}, \mathbf{q}) \cdot \boldsymbol{\phi}_e^*(\mathbf{q})] \, dq = 0; \qquad \mathbf{p} \in B_i. \tag{6.3.11}$$

6.4 Arbitrariness in Somigliana's formula

If $\boldsymbol{\psi}$ is an arbitrary regular displacement vector in B_e, it satisfies the exterior reciprocal relation

$$\int_{\partial B} [\mathbf{g}(\mathbf{p}, \mathbf{q})_e^* \cdot \boldsymbol{\psi}(\mathbf{q}) - \mathbf{g}(\mathbf{p}, \mathbf{q}) \cdot \boldsymbol{\psi}_e^*(\mathbf{q})] \, dq = 0; \qquad \mathbf{p} \in B_i. \tag{6.4.1}$$

Superposing this on Somigliana's interior formula (6.3.1) gives the more general representation formula

$$\int_{\partial B} \mathbf{g}(\mathbf{p}, \mathbf{q})_i^* \cdot [\boldsymbol{\phi}(\mathbf{q}) - \boldsymbol{\psi}(\mathbf{q})] \, dq - \int_{\partial B} \mathbf{g}(\mathbf{p}, \mathbf{q}) \cdot [\boldsymbol{\phi}_i^*(\mathbf{q}) + \boldsymbol{\psi}_e^*(\mathbf{q})] \, dq$$

$$= 4\pi\boldsymbol{\phi}(\mathbf{p}); \qquad \mathbf{p} \in B_i, \tag{6.4.2}$$

D

bearing in mind $\mathbf{g}_i^* + \mathbf{g}_e^* = 0$. Assuming that ϕ is known in $B_i + \partial B$, i.e. ϕ and ϕ_i^* have known continuous values over ∂B, two distinct possibilities arise for ψ:

(i) If $\psi = \phi$ over ∂B, formula (6.4.2) reduces to

$$- \int_{\partial B} \mathbf{g}(\mathbf{p}, \mathbf{q}) \cdot [\phi_i^*(\mathbf{q}) + \psi_e^*(\mathbf{q})] \, dq = 4\pi\phi(\mathbf{p}); \qquad \mathbf{p} \in B_i, \qquad (6.4.3)$$

showing that ϕ may always be generated as a vector simple-layer potential defined by the source density $\sigma = (-1/4\pi)(\phi_i^* + \psi_e^*)$. This representation has already been introduced directly in Chapter 5, and it holds for $\mathbf{p} \in \partial B$ as well as $\mathbf{p} \in B_i$.

(ii) If $\psi_e^* = -\phi_i^*$ over ∂B, formula (6.4.2) reduces to

$$\int_{\partial B} \mathbf{g}(\mathbf{p}, \mathbf{q})_i^* \cdot [\phi(\mathbf{q}) - \psi(\mathbf{q})] \, dq = 4\pi\phi(\mathbf{p}); \qquad \mathbf{p} \in B_i, \qquad (6.4.4)$$

showing that ϕ may always be generated as a vector double-layer potential defined by the source density $\mu = (1/4\pi)(\phi - \psi)$. This representation has also been introduced directly in Chapter 5.

Possibility (i) hinges upon the existence of a unique regular ψ in B_e subject to ψ having prescribed continuous values over ∂B. Similarly, possibility (ii) hinges upon the existence of a unique regular ψ in B_e subject to ψ_e^* having prescribed continuous values over ∂B. These are fundamental existence-uniqueness theorems which will be discussed in Chapter 7.

Formula (6.4.2) has an immediate application to elastic inclusion problems (Eshelby, 1957; Jaswon and Bhargava, 1961). The inclusion is a definite region B_i of an infinite homogeneous elastic continuum, which region undergoes a prescribed irreversible deformation $^\circ\phi$. This takes place against the elastic constraints of the surrounding matrix (i.e. the elastic material occupying B_e), so that the boundary ∂B of B_i suffers an elastic displacement ϕ relative to its irreversible displacement $^\circ\phi$ (Fig. 6.4.1). Hence the net displacement of ∂B is $\phi + {}^\circ\phi$. Now ∂B forms the internal boundary of B_e as well as the (external) boundary of B_i, and from this point of view it suffers a purely elastic displacement $\psi = \phi + {}^\circ\phi$, in order to maintain continuity of contact between B_i and B_e. Neither ϕ nor ψ are known separately, but $\phi - \psi = -{}^\circ\phi$. Under equilibrium conditions, $\phi_i^* + \psi_e^* = 0$ over ∂B, in

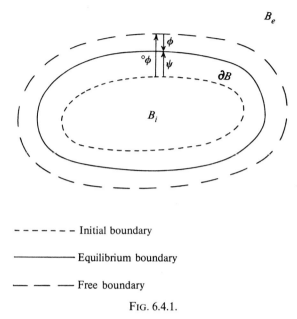

- - - - - - - Initial boundary

――――――― Equilibrium boundary

― ― ― Free boundary

FIG. 6.4.1.

which case (6.4.4) becomes

$$-\int_{\partial B} \mathbf{g}(\mathbf{p}, \mathbf{q})_i^* \cdot {}^\circ\boldsymbol{\phi}(\mathbf{q}) \, \mathrm{d}q = 4\pi\boldsymbol{\phi}(\mathbf{p}); \qquad \mathbf{p} \in B_i. \qquad (6.4.5)$$

Therefore $4\pi\boldsymbol{\phi}$ in B_i may be generated as a vector double-layer potential defined by a known source density $\boldsymbol{\mu} = -{}^\cup\boldsymbol{\phi}$ over ∂B. As $\mathbf{p}$ passes from B_i into B_e, the integral jumps by a total amount $4\pi{}^\circ\boldsymbol{\phi}(\mathbf{p})$ so yielding the value $4\pi(\boldsymbol{\phi} + {}^\circ\boldsymbol{\phi}) = 4\pi\boldsymbol{\psi}$ just inside B_e. We infer that

$$-\int_{\partial B} \mathbf{g}(\mathbf{p}, \mathbf{q})_i^* \cdot {}^\circ\boldsymbol{\phi}(\mathbf{q}) \, \mathrm{d}q = 4\pi\boldsymbol{\psi}(\mathbf{p}); \qquad \mathbf{p} \in B_e, \qquad (6.4.6)$$

which shows that $4\pi\boldsymbol{\psi}$ in B_e may be generated as the exterior counterpart of $4\pi\boldsymbol{\phi}$ in B_i.

The double-layer potentials may be readily converted into simple-layer potentials, and these have the advantage of remaining continuous at ∂B.

Thus, (6.4.5) transforms into

$$- \int_{\partial B} \mathbf{g}(\mathbf{p}, \mathbf{q}) \cdot {}^{\circ}\boldsymbol{\phi}_i^*(\mathbf{q}) \, dq - 4\pi {}^{\circ}\boldsymbol{\phi}(\mathbf{p}) = 4\pi \boldsymbol{\phi}(\mathbf{p}); \qquad \mathbf{p} \in B_i + \partial B,$$

i.e.

$$- \int_{\partial B} \mathbf{g}(\mathbf{p}, \mathbf{q}) \cdot {}^{\circ}\boldsymbol{\phi}_i^*(\mathbf{q}) \, dq = 4\pi [\boldsymbol{\phi}(\mathbf{p}) + {}^{\circ}\boldsymbol{\phi}(\mathbf{p})]; \qquad \mathbf{p} \in B_i + \partial B, \quad (6.4.7)$$

by virtue of Somigliana's formula (6.3.1) applied to ${}^{\circ}\boldsymbol{\phi}$. Similarly, (6.4.6) transforms into

$$- \int_{\partial B} \mathbf{g}(\mathbf{p}, \mathbf{q}) \cdot {}^{\circ}\boldsymbol{\phi}_i^*(\mathbf{q}) \, dq = 4\pi \boldsymbol{\psi}(\mathbf{p}); \qquad \mathbf{p} \in B_e + \partial B, \qquad (6.4.8)$$

by virtue of the *interior* reciprocal relation (6.3.8). Therefore the net displacement vector in $B_i + \partial B$, and the elastic displacement vector in $B_e + \partial B$, may be generated as vector simple-layer potentials defined by

$$\boldsymbol{\sigma} = -\frac{1}{4\pi} {}^{\circ}\boldsymbol{\phi}_i^* = \frac{1}{4\pi} {}^{\circ}\boldsymbol{\phi}_e^*$$

over ∂B. This provides the mathematical justification for Eshelby's direct approach to the problem based upon intuitive physical considerations. A somewhat similar mathematical justification has been presented by Maiti and Makan (1973). A further analysis of elastic inclusions will appear in Chapter 7.

7

Vector Boundary Integral Equations

7.1 Fundamental existence-uniqueness theorems

The fundamental existence-uniqueness theorems of classical elastostatics (Knops and Payne, 1971) closely parallel the corresponding theorems of harmonic function theory. Thus, given arbitrary continuous values of ϕ over ∂B, there exists a unique ϕ in B_i which assumes these boundary values. This is the interior Dirichlet existence-uniqueness theorem for elastostatics. Also, given arbitrary continuous values of ϕ_i^* over ∂B, which satisfy the equilibrium requirements

$$\int_{\partial B} \phi_i^*(\mathbf{p}) \cdot \mathbf{a} \, dp = 0, \qquad \int_{\partial B} \mathbf{p} \wedge \phi_i^*(\mathbf{p}) \cdot \mathbf{b} \, dp = 0, \qquad (7.1.1)$$

where $\mathbf{a}, \mathbf{b}$ are arbitrary constant vectors, there exists a unique (up to an arbitrary rigid-body displacement) ϕ in B_i for which the boundary tractions assume these values. This is the interior Neumann existence-uniqueness theorem for elastostatics.

These theorems may be extended to the infinite domain B_e exterior to ∂B, provided that ϕ is regular at infinity. Thus, given arbitrary continuous values of ϕ over ∂B, there exists a unique regular ϕ in B_e which assumes these boundary values (exterior Dirichlet existence-uniqueness theorem for elastostatics). Also, given arbitrary continuous values of ϕ_e^* over ∂B, there exists a unique regular ϕ in B_e for which the boundary tractions assume these values (exterior Neumann existence-uniqueness theorem for elastostatics). The equilibrium requirements

$$\int_{\partial B} \phi_e^*(\mathbf{p}) \cdot \mathbf{a} \, dp = 0, \qquad \int_{\partial B} \mathbf{p} \wedge \phi_e^*(\mathbf{p}) \cdot \mathbf{b} \, dp = 0 \qquad (7.1.2)$$

need not be satisfied by ϕ_e^*, because of possible compensating effects at infinity following a close analogy with (2.1.1).

The Dirichlet and Neumann boundary conditions of elastostatics are particular cases of a prescribed linear relation

$$\alpha \phi + \beta \phi^* = \mathbf{w} \qquad (7.1.3)$$

between ϕ and ϕ^* at each point of ∂B. Other important cases of (7.1.3) are the Robin problem defined by

$$\alpha < 0, \qquad \beta = 1; \qquad \alpha, \mathbf{w} \in H(\partial B) \qquad (7.1.4)$$

and the mixed problem defined by

$$\left. \begin{array}{l} \left. \begin{array}{l} \alpha = 1 \\ \beta = 0 \end{array} \right\} \text{over } \partial B_1; \qquad \left. \begin{array}{l} \alpha = 0 \\ \beta = 1 \end{array} \right\} \text{over } \partial B_2; \qquad \mathbf{w} \in H(\partial B) \\ \\ \partial B = \partial B_1 + \partial B_2. \end{array} \right\} \qquad (7.1.5)$$

These problems are of considerable engineering significance, but their theory has not yet been completely developed.

To each existence theorem there corresponds a boundary-value problem, viz. that of constructing the appropriate ϕ in $B + \partial B$. A promising approach is to exploit Somigliana's boundary formula (6.3.5) or (6.3.10) which provides a functional constraint between ϕ and ϕ^* over ∂B. Coupling this global constraint with the prescribed local constraint (7.1.3), we have in principle sufficient equations to determine ϕ and ϕ^* over ∂B, whence ϕ may be generated throughout B by the continuation formula (6.3.1) or (6.3.9). This approach may be effectively developed in the Dirichlet and Neumann problems, where it yields vector integral equations analogous to those of the scalar theory. By contrast with (2.2.1), the kernels occurring in these vector equations are strongly singular. However the equations may be transformed into equivalent weakly singular equations by the application of a regularisation operator, thereby allowing a vector Fredholm theory to be implemented.

A complete treatment is available for the isotropic case (Mikhlin, 1965). According to arguments advanced in the text, on the basis of Somigliana's formula, similar properties hold for the general anisotropic case. This approach has been utilised to achieve numerical solutions of some isotropic problems (Cruse, 1969 et seq.) and also of some anisotropic problems (Vogel and Rizzo, 1973). A comparable approach is to represent ϕ by a vector simple-layer or double-layer potential, generated by a hypothetical continuous source distribution over ∂B. The vector potential will be identical with ϕ in B, though not necessarily in $B + \partial B$, provided it satisfies the boundary conditions imposed upon ϕ. This procedure yields vector integral equations, having an adjoint character to those arising from Somigliana's boundary formula, for determining the source densities concerned. As far as we are aware, this approach has been little utilised in practice.

7.2 Rigid-body displacements

A vector field of the form $\phi(\mathbf{p}) = \mathbf{a} + \mathbf{b} \wedge \mathbf{p}$, where $\mathbf{a}, \mathbf{b}$ are constant vectors, defines a rigid-body displacement. This has the property

$$\phi_{\alpha/\beta} + \phi_{\beta/\alpha} = 0; \quad \left.\begin{array}{l} \mathbf{p} \in B_i \\ \alpha, \beta = 1, 2, 3 \end{array}\right\}, \tag{7.2.1}$$

with the accompanying properties $\Phi_{\alpha\beta} = 0$ in B_i and $\phi_i^* = 0$ over ∂B. Conversely, if ϕ satisfies equation (7.2.1), we may infer that $\phi = \mathbf{a} + \mathbf{b} \wedge \mathbf{p}$ where $\mathbf{a}, \mathbf{b}$ are arbitrary constant vectors. This result, combined with (6.2.14), enables us to prove the uniqueness (though not the existence) theorems of the previous section. Thus, $\phi = 0$ over ∂B makes the boundary integral in (6.2.14) vanish, which implies that the strain-energy density $\phi_{\eta/\varepsilon}\Phi_{\eta\varepsilon} = 0$ in B_i since $\phi_{\eta/\varepsilon}\Phi_{\eta\varepsilon} \geqslant 0$ for any stable elastic material. Now the strain-energy density cannot vanish unless all the strain components vanish (see Appendix 7), which implies $\phi = \mathbf{a} + \mathbf{b} \wedge \mathbf{p}$ in B_i since ϕ satisfies (7.2.1). We must choose $\mathbf{a} = \mathbf{b} = 0$ in order to ensure the continuity of ϕ in B_i with $\phi = 0$ over ∂B, so proving that the interior Dirichlet problem has a unique solution. Similarly, $\phi_i^* = 0$ over ∂B implies $\phi = \mathbf{a} + \mathbf{b} \wedge \mathbf{p}$ in $B_i + \partial B$, so proving that the interior Neumann problem has a unique solution up to an arbitrary rigid-body displacement. There is no difficulty in proving that the corresponding exterior problems have unique regular solutions, utilising the exterior version of (6.2.14). Straightforward adaptations of the argument prove that the boundary-value problems (7.1.4), (7.1.5) have unique solutions.

According to the interior Dirichlet existence-uniqueness theorem, $\boldsymbol{\phi} = \mathbf{a} + \mathbf{b} \wedge \mathbf{p}$ over ∂B implies $\boldsymbol{\phi} = \mathbf{a} + \mathbf{b} \wedge \mathbf{p}$ in B_i. Also, as has been noted, $\boldsymbol{\phi}_i^* = 0$ over ∂B implies an arbitrary $\boldsymbol{\phi} = \mathbf{a} + \mathbf{b} \wedge \mathbf{p}$ in $B_i + \partial B$. Clearly, therefore, the rigid-body displacement vector plays an analogous role to the constant in scalar potential theory. This analogy offers an extremely useful guide to the analysis of vector integral equations associated with elastostatics.

It is convenient to break down $\mathbf{a} + \mathbf{b} \wedge \mathbf{p}$ into the six independent vectors

$$\left. \begin{array}{ccc} \boldsymbol{\mu}_1 = \langle 1,0,0 \rangle, & \boldsymbol{\mu}_2 = \langle 0,1,0 \rangle, & \boldsymbol{\mu}_3 = \langle 0,0,1 \rangle \\ \boldsymbol{\mu}_4 = \langle 1,0,0 \rangle \wedge \mathbf{p}, & \boldsymbol{\mu}_5 = \langle 0,1,0 \rangle \wedge \mathbf{p}, & \boldsymbol{\mu}_6 = \langle 0,0,1 \rangle \wedge \mathbf{p} \end{array} \right\}, \quad (7.2.2)$$

each of which satisfies the homogeneous equation

$$\int_{\partial B} \mathbf{g}(\mathbf{p}, \mathbf{q})_i^* \cdot \boldsymbol{\mu}(\mathbf{q}) \, dq - 2\pi\boldsymbol{\mu}(\mathbf{p}) = 0; \qquad \mathbf{p} \in \partial B. \qquad (7.2.3)$$

This can be readily seen by putting $\boldsymbol{\phi} = \boldsymbol{\mu}_s, \boldsymbol{\phi}_i^* = 0;\ s = 1,2,\ldots,6$ into Somigliana's boundary formula (6.3.5), or directly from (5.4.16), (5.4.18). The adjoint of equation (7.2.3) is

$$\int_{\partial B} \mathbf{g}_i^*(\mathbf{p}, \mathbf{q}) \cdot \boldsymbol{\lambda}(\mathbf{q}) \, dq - 2\pi\boldsymbol{\lambda}(\mathbf{p}) = 0; \qquad \mathbf{p} \in \partial B, \qquad (7.2.4)$$

which has six independent non-trivial solutions $\boldsymbol{\lambda}_s;\ s = 1,2,\ldots,6$ uniquely defined by the relations

$$\int_{\partial B} \mathbf{g}(\mathbf{p}, \mathbf{q}) \cdot \boldsymbol{\lambda}_s(\mathbf{q}) \, dq = \boldsymbol{\mu}_s(\mathbf{p}); \qquad \left. \begin{array}{c} \mathbf{p} \in \partial B \\ s = 1,2,\ldots,6 \end{array} \right\}. \qquad (7.2.5)$$

We shall refer to $\boldsymbol{\lambda}_s;\ s = 1,2,\ldots,6$ as the normalised non-trivial solutions of equations (7.2.4). Thus, putting $\boldsymbol{\phi} = \boldsymbol{\mu}_s, \boldsymbol{\phi}_i^* = 0$ into the general representation formula (6.4.3), we obtain

$$-\frac{1}{4\pi} \int_{\partial B} \mathbf{g}(\mathbf{p}, \mathbf{q}) \cdot \boldsymbol{\psi}_e^*(\mathbf{q}) \, dq = \boldsymbol{\mu}_s(\mathbf{p}); \qquad \mathbf{p} \in B_i, \qquad (7.2.6)$$

where $\boldsymbol{\psi}$ is the unique regular field which exists in B_e subject to $\boldsymbol{\psi} = \boldsymbol{\mu}_s$ over

∂B. Since the left-hand side of (7.2.6) remains continuous at ∂B, this represesentation remains valid at ∂B and we infer that $\lambda_s = -(1/4\pi)\psi_e^*$. Relation (7.2.5) may be regarded as an integral equation for λ_s in terms of μ_s, with equation (7.2.4) as the accompaning traction equation.

A Volterra dislocation is a plane or curved cut ∂B within the continuum, across which the elastostatic displacement vector jumps by a prescribed amount $\mathbf{a} + \mathbf{b} \wedge \mathbf{p}$, the tractions remaining continuous (Nabarro, 1967). We may regard this as a disc-like inclusion of vanishing thickness, bounded by surfaces $\partial B_+, \partial B_-$ which suffer the irreversible displacements

$$^{\circ}\boldsymbol{\phi}_+ = \tfrac{1}{2}(\mathbf{a} + \mathbf{b} \wedge \mathbf{p}), \qquad ^{\circ}\boldsymbol{\phi}_- = -\tfrac{1}{2}(\mathbf{a} + \mathbf{b} \wedge \mathbf{p})$$

respectively (Fig. 7.2.1). Substituting these into (6.4.6) and bearing in mind

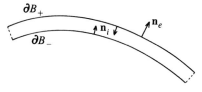

$$\partial B_+$$
$$\mathbf{n}_i \qquad \mathbf{n}_e$$
$$\partial B_-$$

FIG. 7.2.1

$\partial B = \partial B_- + \partial B_+$, we obtain

$$4\pi\psi(\mathbf{p}) = -\tfrac{1}{2}\int_{\partial B_+} \mathbf{g}(\mathbf{p},\mathbf{q})_i^* \cdot (\mathbf{a} + \mathbf{b} \wedge \mathbf{q}) \, dq + \tfrac{1}{2}\int_{\partial B_-} \mathbf{g}(\mathbf{p},\mathbf{q})_i^* \cdot (\mathbf{a} + \mathbf{b} \wedge \mathbf{q}) \, dq$$

$$- \int_{\partial B_+} \mathbf{g}(\mathbf{p},\mathbf{q})_e^* \cdot (\mathbf{a} + \mathbf{b} \wedge \mathbf{q}) \, dq; \qquad \mathbf{p} \in B_e. \qquad (7.2.7)$$

The dislocation is therefore mathematically equivalent to a distribution of vector double sources over ∂B_+, of density

$$\mu = \frac{1}{4\pi}(\mathbf{a} + \mathbf{b} \wedge \mathbf{q}) \qquad \text{at} \qquad \mathbf{q} \in \partial B_+,$$

being therefore the analogue of the vortex sheet or magnetic shell of scalar potential theory (Section 1.6). This can also be seen directly from the discontinuity properties of vector double-layer potentials (Section 5.4). By virtue of (7.2.2), there are six independent Volterra dislocations, each of which may have two senses.

7.3 Representation by vector simple-layer potentials

Given continuous values of ϕ over ∂B, there exist a unique ϕ in B_i and a unique regular ψ in B_e which assume these values. Accordingly, ϕ_i^* and ψ_e^* are available in principle over ∂B, so allowing us to utilise the representation (6.4.3) for ϕ, viz.

$$\int_{\partial B} \mathbf{g}(\mathbf{p}, \mathbf{q}) . \sigma(\mathbf{q}) \, dq = \phi(\mathbf{p}); \qquad \mathbf{p} \in B_i \qquad (7.3.1)$$

where $\sigma = -\dfrac{1}{4\pi}(\phi_i^* + \psi_e^*)$. This representation has been introduced already, but without any implication for boundary-value problems. Since the integral remains continuous at ∂B, the representation remains valid at ∂B and yields the boundary relation

$$\int_{\partial B} \mathbf{g}(\mathbf{p}, \mathbf{q}) . \sigma(\mathbf{q}) \, dq = \phi(\mathbf{p}); \qquad \mathbf{p} \in \partial B. \qquad (7.3.2)$$

This may be regarded as a vector integral equation for σ in terms of ϕ over ∂B. It has a unique solution which enables us to generate ϕ throughout B_i from (7.3.1) thereby solving the interior Dirichlet problem for any closed Liapunov surface.

Subject to $\phi = O(r^{-1})$ as $r \to \infty$, the exterior Dirichlet problem of elastostatics is also solved by equation (7.3.2). Operating upon the left-hand side of it by $\int_{\partial B} \lambda_s(\mathbf{p}) . (\ldots) \, dp$, we find

$$\int_{\partial B} \lambda_s(\mathbf{p}) . \left\{ \int_{\partial B} \mathbf{g}(\mathbf{p}, \mathbf{q}) . \sigma(\mathbf{q}) \, dq \right\} dp = \int_{\partial B} \left\{ \int_{\partial B} \lambda_s(\mathbf{p}) . \mathbf{g}(\mathbf{p}, \mathbf{q}) \, dp \right\} . \sigma(\mathbf{q}) \, dq$$

$$= \int_{\partial B} \mu_s(\mathbf{q}) . \sigma(\mathbf{q}) \, dq, \qquad (7.3.3)$$

assuming that the order of integration may be interchanged and using an obvious adaptation of (7.2.5). It follows that

$$\int_{\partial B} \sigma(\mathbf{q}) . \mu_s(\mathbf{q}) \, dq = \int_{\partial B} \phi(\mathbf{p}) . \lambda_s(\mathbf{p}) \, dp; \qquad s = 1, 2, , \ldots, 6. \qquad (7.3.4)$$

Reference to (6.1.4), (6.1.5) shows that

$$4\pi \cdot \int_{\partial B} \sigma(\mathbf{q}) \cdot \mu_s(\mathbf{q}) \, dq, \qquad \text{i.e. } 4\pi \int_{\partial B} \phi(\mathbf{p}) \cdot \lambda_s(\mathbf{p}) \, dp; \qquad s = 1, 2, 3 \qquad (7.3.5)$$

$$4\pi \int_{\partial B} \sigma(\mathbf{q}) \cdot \mu_s(\mathbf{q}) \, dq, \qquad \text{i.e. } 4\pi \int_{\partial B} \phi(\mathbf{p}) \cdot \lambda_s(\mathbf{p}) \, dp; \qquad s = 4, 5, 6 \qquad (7.3.6)$$

define the components of the force and moment resultants, respectively, arising from the tractions at infinity generated by σ over ∂B.

Given continuous values of ϕ_i^* over ∂B, there exists an appropriate ϕ in B_i for which the representation (7.3.1) may again be utilised. Clearly σ satisfies the interior traction equation

$$\int_{\partial B} \mathbf{g}_i^*(\mathbf{p}, \mathbf{q}) \cdot \sigma(\mathbf{q}) \, dq - 2\pi\sigma(\mathbf{p}) = \phi_i^*(\mathbf{p}); \qquad \mathbf{p} \in \partial B, \qquad (7.3.7)$$

of which the special case $\phi_i^* = 0$ has already been examined. Equation (7.3.7) has a solution only if

$$\int_{\partial B} \phi_i^*(\mathbf{p}) \cdot \mu_s(\mathbf{p}) \, dp = 0; \qquad s = 1, 2, \ldots, 6, \qquad (7.3.8)$$

as may be proved by operating upon both sides of it by $\int_{\partial B} \mu_s(\mathbf{p}) \cdot (\ldots) \, dp$. Conditions (7.3.8) are equivalent to the previously stated equilibrium conditions (7.1.1). Once a particular solution σ_0 of equation (7.3.7) has been achieved, the general solution appears as

$$\sigma = \sigma_0 + \sum_{s=1}^{6} k_s \lambda_s \qquad (7.3.9)$$

where the k_s are arbitrary constants. This generates the class of displacements

$$\int_{\partial B} \mathbf{g}(\mathbf{p}, \mathbf{q}) \cdot \sigma_0(\mathbf{q}) \, dq + \sum_{s=1}^{6} k_s \mu_s; \qquad \mathbf{p} \in B_i + \partial B, \qquad (7.3.10)$$

each characterised by the same ϕ_i^* over ∂B.

The exterior traction equation

$$\int_{\partial B} \mathbf{g}_e^*(\mathbf{p}, \mathbf{q}) \cdot \sigma(\mathbf{q}) \, dq - 2\pi\sigma(\mathbf{p}) = \phi_e^*(\mathbf{p}); \qquad \mathbf{p} \in \partial B \qquad (7.3.11)$$

has a homogeneous component equation

$$\int_{\partial B} \mathbf{g}_e^*(\mathbf{p}, \mathbf{q}) . \sigma(\mathbf{q}) \, dq - 2\pi\sigma(\mathbf{p}) = 0; \qquad \mathbf{p} \in \partial B, \qquad (7.3.12)$$

with an accompanying adjoint equation

$$\int_{\partial B} \mathbf{g}(\mathbf{p}, \mathbf{q})_e^* . \tau(\mathbf{q}) \, dq - 2\pi\tau(\mathbf{p}) = 0; \qquad \mathbf{p} \in \partial B. \qquad (7.3.13)$$

It may be proved independently that neither of these homogeneous equations admits any non-trivial solution. The proof for equation (7.3.12) follows essentially the same lines as that for the corresponding scalar equation (2.4.8). The proof for equation (7.3.13) employs Somigliana's exterior boundary formula (6.3.10). Thus, choosing $\phi = \tau$ over ∂B, we infer from (6.3.10) that $\phi_e^* = 0$ over ∂B, which implies that $\phi = 0$ in $B_e + \partial B$ whence $\tau = 0$ over ∂B. It follows on general grounds that equation (7.3.11) has a unique continuous solution for arbitrary continuous values of ϕ_e^* over ∂B. Operating upon each side of it by $\int_{\partial B} \mu_s(\mathbf{p}) . (\dots) \, dp$, we arrive without difficulty at the integral relations

$$-4\pi \int_{\partial B} \sigma(\mathbf{q}) . \mu_s(\mathbf{q}) \, dq = \int_{\partial B} \phi_e^*(\mathbf{p}) . \mu_s(\mathbf{p}) \, dp; \qquad s = 1, 2, \dots, 6, \qquad (7.3.14)$$

which form the Neumann counterpart of the Dirichlet integral relations (7.3.4). Since the right-hand side of (7.3.14) defines the components of the force and moment resultants over ∂B associated with ϕ_e^*, this equation verifies directly that (7.3.5), (7.3.6) define the compensating resultants generated at infinity.

7.4 Representation by vector double-layer potentials

The interior Dirichlet problem may also be formulated by introducing the representation

$$\int_{\partial B} \mathbf{g}(\mathbf{p}, \mathbf{q})_i^* . \mu(\mathbf{q}) \, dq = \phi(\mathbf{p}); \qquad \mathbf{p} \in B_i \qquad (7.4.1)$$

for ϕ, where $\mu = \frac{1}{4\pi}(\phi - \psi)$. As in (6.4.4), ψ is the unique regular exterior field subject to $\psi_e^* = -\phi_i^*$ over ∂B, where ϕ_i^* is available in principle since ϕ is available in principle in $B_i + \partial B$. The integral jumps by $-2\pi\mu(\mathbf{p})$ as $\mathbf{p}$ passes from B_i onto ∂B, so yielding the boundary relation

$$\int_{\partial B} \mathbf{g}(\mathbf{p}, \mathbf{q})_i^* \cdot \mu(\mathbf{q})\, dq + 2\pi\mu(\mathbf{p}) = \phi(\mathbf{p}); \qquad \mathbf{p} \in \partial B. \qquad (7.4.2)$$

This may be regarded as a vector integral equation for μ in terms of ϕ over ∂B. Its homogeneous component is mathematically equivalent to equation (7.3.12), which has no non-trivial solutions. Consequently equation (7.4.2) has a unique solution for arbitrary continuous values of ϕ over ∂B. For example, if $\phi(\mathbf{p}) = \mathbf{a} + \mathbf{b} \wedge \mathbf{p}$ over ∂B, then $\mu(\mathbf{p}) = \frac{1}{4\pi}(\mathbf{a} + \mathbf{b} \wedge \mathbf{p})$.

The exterior Dirichlet problem cannot in general be formulated by a double-layer representation. This is essentially because of the null force and moment resultants characteristic of $\mathbf{W}$, as expressed by (6.1.19), (6.1.20). Alternatively, the exterior equation

$$\int_{\partial B} \mathbf{g}(\mathbf{p}, \mathbf{q})_e^* \cdot \mu(\mathbf{q})\, dq + 2\pi\mu(\mathbf{p}) = \phi(\mathbf{p}); \qquad \mathbf{p} \in \partial B \qquad (7.4.3)$$

has a homogeneous component which is mathematically equivalent to equation (7.2.3). This latter has six non-trivial solutions μ_s; $s = 1, 2, \ldots, 6$. Therefore equation (7.4.3) only admits a solution if

$$\int_{\partial B} \phi(\mathbf{p}) \cdot \lambda_s(\mathbf{p})\, dp = 0; \qquad s = 1, 2, \ldots, 6, \qquad (7.4.4)$$

where λ_s; $s = 1, 2, \ldots, 6$ are the normalised non-trivial solutions of the adjoint equation (7.2.4). Reference to (7.3.5), (7.3.6) shows that conditions (7.4.4) express the requirement for null force and moment resultants over ∂B. An extension of the exterior double-layer representation, analogous to the extended scalar representation (2.6.6), has been given by Watson (1972a).

It may be remarked that the representation (7.4.1) is mathematically equivalent to the displacement generated by a continuous distribution of dislocations over ∂B, as follows from the definition of a Volterra dislocation given in Section 7.2. This point of view has been taken intuitively by Lardner (1972) in solving certain two-dimensional problems. A mathematical justification for Lardner's approach, based upon Somigliana's formula, has been given by Palit (1976). The equivalence between static dislocations and inclusions (Section 6.4) has been established by Willis (1965).

7.5 Representation by Somigliana's formula

Given ϕ over ∂B, formula (6.3.5) yields the vector integral equation

$$\int_{\partial B} \mathbf{g}(\mathbf{p}, \mathbf{q}) \cdot \phi_i^*(\mathbf{q})\, dq = \int_{\partial B} \mathbf{g}(\mathbf{p}, \mathbf{q})_i^* \cdot \phi(\mathbf{q})\, dq - 2\pi\phi(\mathbf{p}); \qquad \mathbf{p} \in \partial B, \qquad (7.5.1)$$

for ϕ_i^* over ∂B. This is of the form (7.3.1) and it therefore has a unique solution, which identically satisfies the equilibrium conditions (7.3.8). Thus, operating upon the left-hand side by $\int_{\partial B} \lambda_s(\mathbf{p}) \cdot (\ldots)\, dp$, we find

$$\int_{\partial B} \lambda_s(\mathbf{p}) \cdot \left\{ \int_{\partial B} \mathbf{g}(\mathbf{p}, \mathbf{q}) \cdot \phi_i^*(\mathbf{q})\, dq \right\} dp = \int_{\partial B} \left\{ \int_{\partial B} \lambda_s(\mathbf{p}) \cdot \mathbf{g}(\mathbf{p}, \mathbf{q})\, dp \right\} \cdot \phi_i^*(\mathbf{q})\, dq$$

$$= \int_{\partial B} \mu_s(\mathbf{q}) \cdot \phi_i^*(\mathbf{q})\, dq,$$

assuming that the order of integration may be interchanged and using an obvious adaptation of (7.2.5). Also, operating upon the right-hand integral by $\int_{\partial B} \lambda_s(\mathbf{p}) \cdot (\ldots)\, dp$, we find

$$\int_{\partial B} \lambda_s(\mathbf{p}) \cdot \left\{ \int_{\partial B} \mathbf{g}(\mathbf{p}, \mathbf{q})_i^* \cdot \phi(\mathbf{q})\, dq \right\} dp = \int_{\partial B} \left\{ \int_{\partial B} \lambda_s(\mathbf{p}) \cdot \mathbf{g}(\mathbf{p}, \mathbf{q})_i^*\, dp \right\} \cdot \phi(\mathbf{q})\, dq$$

$$= 2\pi \int_{\partial B} \lambda_s(\mathbf{q}) \cdot \phi(\mathbf{q})\, dq,$$

assuming that the order of integration may be interchanged and using an obvious adaptation of (7.2.4). It follows that

$$\int_{\partial B} \mu_s(\mathbf{q}) \cdot \phi_i^*(\mathbf{q})\, dq = 2\pi \int_{\partial B} \lambda_s(\mathbf{q}) \cdot \phi(\mathbf{q})\, dq - 2\pi \int_{\partial B} \lambda_s(\mathbf{p}) \cdot \phi(\mathbf{p})\, dp = 0$$

in accordance with (7.3.8).

Given ϕ_i^* over ∂B, formula (6.3.5) yields the vector integral equation

$$\int_{\partial B} \mathbf{g}(\mathbf{p}, \mathbf{q})_i^* \cdot \phi(\mathbf{q}) \, dq - 2\pi\phi(\mathbf{p}) = \int_{\partial B} \mathbf{g}(\mathbf{p}, \mathbf{q}) \cdot \phi_i^*(\mathbf{q}) \, dq; \qquad \mathbf{p} \in \partial B, \tag{7.5.2}$$

for ϕ over ∂B. This is mathematically equivalent to an equation of the form (7.4.3), and it therefore only admits a solution subject to the conditions

$$\int_{\partial B} \lambda_s(\mathbf{p}) \cdot \left\{ \int_{\partial B} \mathbf{g}(\mathbf{p}, \mathbf{q}) \cdot \phi_i^*(\mathbf{q}) \, dq \right\} dp = 0; \qquad s = 1, 2, \ldots, 6,$$

i.e. $$\int_{\partial B} \mu_s(\mathbf{q}) \cdot \phi_i^*(\mathbf{q}) \, dq = 0; \qquad s = 1, 2, \ldots, 6,$$

which are the expected equilibrium conditions. Once a particular solution ϕ_0 has been achieved, the general solution appears as

$$\phi = \phi_0 + \sum_{s=1}^{6} k_s \lambda_s. \tag{7.5.3}$$

For instance, if $\phi_i^* = 0$ over ∂B, then $\phi = \sum_{s=1}^{6} k_s \lambda_s$. Introducing this solution into the left-hand side of Somigliana's formula (6.3.1), we generate $4\pi \sum_{s=1}^{6} k_s \lambda_s$ in B_i.

The exterior Dirichlet problem is formulated by the vector integral equation

$$\int_{\partial B} \mathbf{g}(\mathbf{p}, \mathbf{q}) \cdot \phi_e^*(\mathbf{q}) \, dq = \int_{\partial B} \mathbf{g}(\mathbf{p}, \mathbf{q})_e^* \cdot \phi(\mathbf{q}) \, dq - 2\pi\phi(\mathbf{p}); \qquad \mathbf{p} \in \partial B, \tag{7.5.4}$$

for ϕ_e^* over ∂B. This has a unique solution, which satisfies the integral relations

$$\int_{\partial B} \mu_s(\mathbf{q}) \cdot \phi_e^*(\mathbf{q}) \, dq = -4\pi \int_{\partial B} \lambda_s(\mathbf{p}) \cdot \phi(\mathbf{p}) \, dp; \qquad s = 1, 2, \ldots, 6, \tag{7.5.5}$$

as may be proved by operating upon both sides by $\int_{\partial B} \lambda_s(\mathbf{p}) \cdot (\ldots) \, dp$. These are the same as those obtained by eliminating σ between (7.3.4) and (7.3.14).

The exterior Neumann problem is formulated by the vector integral equation

$$\int_{\partial B} \mathbf{g}(\mathbf{p}, \mathbf{q})_e^* \cdot \phi(\mathbf{q}) \, dq \; - \; 2\pi\phi(\mathbf{p}) = \int_{\partial B} \mathbf{g}(\mathbf{p}, \mathbf{q}) \cdot \phi_e^*(\mathbf{q}) \, dq; \qquad \mathbf{p} \in \partial B, \qquad (7.5.6)$$

for ϕ over ∂B. This is mathematically equivalent to an equation of the form (7.4.2), and it therefore has a unique solution which satisfies relations (7.5.5).

8

Isotropic Elasticity: Displacement Representations

8.1 Papkovich-Neuber representation

An isotropic linear elastic medium is characterised by the elastic constant tensor

$$C^{\alpha\beta\xi\eta} = \mu(\delta_{\alpha\xi}\delta_{\beta\eta} + \delta_{\alpha\eta}\delta_{\beta\xi}) + \lambda\delta_{\alpha\beta}\delta_{\xi\eta}, \tag{8.1.1}$$

which involves only two independent components. The explicit constants μ, λ are Lamé's elastic constants, and μ may be identified physically as the shear modulus. Introducing (8.1.1) into the stress–strain relations (5.1.3) yields

$$\Phi_{\alpha\beta} = \mu(\phi_{\alpha/\beta} + \phi_{\beta/\alpha}) + \lambda\nabla\cdot\phi\delta_{\alpha\beta}, \tag{8.1.2}$$

where $\nabla\cdot\phi$ ($= \phi_{\alpha/\alpha}$) is the dilatation associated with ϕ. We substitute from (8.1.2) into equation (5.1.5) to obtain the Cauchy–Navier equation

$$\mu\nabla^2\phi + (\lambda + \mu)\nabla(\nabla\cdot\phi) = 0 \tag{8.1.3}$$

governing ϕ in a source-free domain. This is preferably written

$$\nabla^2\phi + \frac{1}{1 - 2v}\nabla(\nabla\cdot\phi) = 0; \quad v = \frac{\lambda}{2(\lambda + \mu)}, \tag{8.1.4}$$

where v is Poisson's ratio ($0 < v \leqslant \frac{1}{2}$).

A general solution of equation (8.1.4) has been given by Papkovich (1932) and Neuber (1934), viz.

$$\phi(\mathbf{p}) = \mathbf{h} - \kappa\nabla(\mathbf{p}.\mathbf{h} + f); \qquad \left.\begin{array}{c} \nabla^2\mathbf{h(p)} = 0, \qquad \nabla^2 f(\mathbf{p}) = 0 \\[2mm] \kappa^{-1} = 4(1 - v) \end{array}\right\} \qquad (8.1.5)$$

where $\mathbf{h}$ is a harmonic vector and f is a harmonic scalar. To prove that (8.1.5) provides a solution for arbitrary $\mathbf{h}$ and f, we note that

$$\nabla^2(\mathbf{p}.\mathbf{h}) = 2\nabla.\mathbf{h} \quad \text{since} \quad \nabla^2\mathbf{h} = 0, \qquad (8.1.6)$$

whence

$$\nabla.\phi = \nabla.\mathbf{h} - \kappa\nabla^2(\mathbf{p}.\mathbf{h} + f) = (1 - 2\kappa)\nabla.\mathbf{h}, \qquad (8.1.7)$$

$$\nabla^2\phi = \nabla^2\mathbf{h} - \kappa\nabla^2\nabla(\mathbf{p}.\mathbf{h} + f)$$

$$= 0 - \kappa\nabla\nabla^2(\mathbf{p}.\mathbf{h} + f) = -2\kappa\nabla(\nabla.\mathbf{h}), \qquad (8.1.8)$$

in which case the left-hand side of equation (8.1.4) becomes

$$-2\kappa\nabla(\nabla.\mathbf{h}) + \frac{1 - 2\kappa}{1 - 2v}\nabla(\nabla.\mathbf{h}) = 0 \quad \text{since} \quad \kappa^{-1} = 4(1 - v).$$

It has been remarked by Mindlin (1936) that the Papkovich–Neuber representation is equivalent to the Galerkin vector representation

$$\phi = \nabla^2\boldsymbol{\chi} - 2\kappa\nabla(\nabla.\boldsymbol{\chi}); \qquad \nabla^4\boldsymbol{\chi} = 0, \qquad (8.1.9)$$

as readily follows by writing $\nabla^2\boldsymbol{\chi} = \mathbf{h}$ and noting that

$$\nabla^2(\nabla.\boldsymbol{\chi}) = \nabla.\mathbf{h} = \tfrac{1}{2}\nabla^2(\mathbf{p}.\mathbf{h}), \qquad \text{i.e. } \nabla.\boldsymbol{\chi} = \tfrac{1}{2}(\mathbf{p}.\mathbf{h} + f); \qquad \nabla^2 f = 0.$$

To demonstrate the generality of (8.1.5), we introduce the Stokes–Helmholtz resolution (Sommerfeld, 1964)

$$\phi = \nabla S + \nabla \wedge \mathbf{A}; \qquad \nabla.\mathbf{A} = 0. \qquad (8.1.10)$$

Substituting this into equation (8.1.4), and noting that $\nabla \cdot \phi = \nabla^2 S$, we find

$$\nabla^2 \left(\nabla S + \nabla \wedge A + \frac{1}{1 - 2\nu} \nabla S \right) = 0, \tag{8.1.11}$$

which implies

$$\frac{2(1 - \nu)}{1 - 2\nu} \nabla S + \nabla \wedge A = h; \qquad \nabla^2 h = 0. \tag{8.1.12}$$

From (8.1.12)

$$\nabla^2 S = \frac{1 - 2\nu}{2(1 - \nu)} \nabla \cdot h = \frac{1 - 2\nu}{4(1 - \nu)} \nabla^2 (p \cdot h)$$

and therefore

$$S = \frac{1 - 2\nu}{4(1 - \nu)} (p \cdot h + f); \qquad \nabla^2 f = 0. \tag{8.1.13}$$

An explicit expression for A is not required, since $\nabla \wedge A$ may be eliminated between (8.1.10) and (8.1.12) to yield

$$\phi = \nabla S + h - \frac{2(1 - \nu)}{1 - 2\nu} \nabla S = h - \frac{1}{1 - 2\nu} \nabla S$$

$$= h - \frac{1}{4(1 - \nu)} \nabla(p \cdot h + f).$$

For an incompressible medium, $\kappa = \nu = \frac{1}{2}$ and the Papkovich–Neuber representation becomes

$$\phi = h - \frac{1}{2} \nabla(p \cdot h + f). \tag{8.1.14}$$

This automatically ensures that $\nabla \cdot \phi = 0$. The ratio $\nabla \cdot \phi / (1 - 2\nu)$ tends to $-P/\mu$ as $\nu \to \frac{1}{2}$, $\nabla \cdot \phi \to 0$, where P is the pressure. Accordingly, equation (8.1.4) transforms into

$$\mu \nabla^2 \phi - \nabla P = 0; \qquad \nabla \cdot \phi = 0, \tag{8.1.15}$$

where P is an unknown to be determined along with ϕ. (An alternative analysis of equation (8.1.15) has been given by Golecki (1974).) A representation for P in parallel with (8.1.14) follows by noting that

$$\nabla P = \mu\nabla^2\phi = \mu\nabla^2\mathbf{h} - \tfrac{1}{2}\mu\nabla\nabla^2(\mathbf{p}.\mathbf{h} + f) = -\mu\nabla(\nabla.\mathbf{h}),$$

whence

$$P = -\mu\nabla.\mathbf{h} + c; \qquad c = \text{an arbitrary constant.} \qquad (8.1.16)$$

This analysis covers the Stokes flow of a slow viscous fluid, on identifying ϕ with the local fluid velocity and P with the local hydrostatic pressure. An interesting choice for h is

$$\mathbf{h} = \frac{\mathbf{F}}{4\pi r}; \qquad r = |\mathbf{p}|,$$

where $\mathbf{F}$ is a constant vector. Inserting this into (8.1.14) yields the Stokeslet solution

$$\phi = \frac{r^2F + (\mathbf{F}.\mathbf{p})\mathbf{p}}{8\pi\mu r^3},$$

which has been utilised by Youngren and Acrivos (1975), Lighthill (1976) and Priyakumari (1977).

It is often possible to dispense with the harmonic function f in (8.1.5). Alternatively expressed, it is often possible to write

$$-\kappa\nabla f = \mathbf{h} - \kappa\nabla(\mathbf{p}.\mathbf{h}); \qquad \nabla^2\mathbf{h} = 0, \qquad \nabla^2 f = 0, \qquad (8.1.17)$$

where $\mathbf{h}$ can be determined in terms of f. This is most easily achieved by putting $\mathbf{h} = \nabla s$, $\nabla^2 s = 0$ and substituting into the right-hand side of equation (8.1.17) to obtain

$$-\kappa\nabla f = \nabla s - \kappa\nabla(\mathbf{p}.\nabla s) = \nabla(s - \kappa\mathbf{p}.\nabla s), \qquad (8.1.18)$$

where the expression in the final brackets is a harmonic function. Equation (8.1.18) yields the equation

$$s - \kappa\mathbf{p}.\nabla s = -\kappa f \qquad (8.1.19)$$

for s in terms of f, which is solved in Appendix 8 for a domain of any shape. To fix ideas, suppose that f is given within a sphere B. Then the convergent expansion

$$f = \sum_{n=0}^{\infty} \sum_{m=-n}^{n} b_n^m r^n S_n^m; \qquad S_n^m = P_n^m(\cos \theta) \, e^{im\psi} \qquad (8.1.20)$$

holds within B, utilising the usual symbolism for spherical harmonic functions, where the b_n^m are uniquely determined coefficients. We assume a corresponding expansion

$$g = \sum_{n=0}^{\infty} \sum_{m=-n}^{n} a_n^m r^n S_n^m \qquad (8.1.21)$$

within B, where the a_n^m are coefficients to be determined in terms of the b_n^m. Substituting (8.1.20), (8.1.21) into (8.1.19), and equating coefficients of $r^n S_n^m$ on each side, we find

$$(1 - \kappa n) a_n^m = -\kappa b_n^m, \qquad (8.1.22)$$

i.e. $$a_n^m = -\kappa b_n^m/(1 - \kappa n). \qquad (8.1.23)$$

Accordingly, unless $1 - \kappa n = 0$, the expansion (8.1.21) certainly converges thereby defining a unique harmonic function s inside B. Clearly however s, and therefore $\mathbf{h}$, may not exist if $\kappa = n^{-1}$, i.e. if $v = 1 - n/4$ where n is any integer, though the only physically admissible possibilities are $n = 2$ (incompressible medium) and $n = 3$. An alternative study has been made by Eubanks and Sternberg (1956).

8.2 Kelvin solution

The vector $\mathbf{h} = \langle |\mathbf{p}|^{-1}, 0, 0 \rangle$ is harmonic everywhere except at $\mathbf{p} = 0$, and therefore

$$\phi = \langle r^{-1}, 0, 0 \rangle - \kappa \nabla(p_1/r); \qquad r = |\mathbf{p}| \qquad (8.2.1)$$

satisfies equation (8.1.4) everywhere except at $\mathbf{p} = 0$. More generally,

$$\phi = \langle R^{-1}, 0, 0 \rangle - \kappa \nabla\left(\frac{p_1 - q_1}{R}\right); \qquad R = |\mathbf{p} - \mathbf{q}| \qquad (8.2.2)$$

satisfies equation (8.1.4) everywhere except at $\mathbf{p} = \mathbf{q}$. This is Kelvin's solution for the displacement vector generated by a concentrated force acting in the 1-direction at $\mathbf{q}$. Analysis (see below) shows that the magnitude of the concentrated force in question is $4\pi\mu$. Analogous expressions to (8.2.2) hold for concentrated forces of magnitude $4\pi\mu$ acting in the 2- and 3-directions at $\mathbf{q}$. Separating into components and utilising the symbolism of (5.2.7), we arrive at the isotropic fundamental displacement dyadic

$$g(\mathbf{p}_\alpha, \mathbf{q}_\eta) = \frac{(1 - \kappa)\delta_{\alpha\eta}}{\mu R} + \frac{\kappa}{\mu R} \frac{\partial R}{\partial p_\alpha} \frac{\partial R}{\partial p_\eta}; \qquad \alpha, \eta = 1, 2, 3, \qquad (8.2.3)$$

where p_α; $\alpha = 1, 2, 3$ are the components of $\mathbf{p}$. This expression defines the displacement component in the α-direction at $\mathbf{p}$ generated by a concentrated force of magnitude 4π acting in the η-direction at $\mathbf{q}$.

It will be seen from (8.2.3) that

$$g(\mathbf{p}_\alpha, \mathbf{q}_\eta) = g(\mathbf{p}_\eta, \mathbf{q}_\alpha) = g(\mathbf{q}_\eta, \mathbf{p}_\alpha). \qquad (8.2.4)$$

The associated fundamental traction dyadics are

$$g^*(\mathbf{p}_\alpha, \mathbf{q}_\eta) = \frac{(2v - 1)}{2(1 - v)} \frac{1}{R^2} \left[\frac{\partial R}{\partial p_\alpha} n_\eta - \frac{\partial R}{\partial p_\eta} n_\alpha \right.$$
$$\left. + \frac{\partial R}{\partial n} \left\{ \delta_{\alpha\eta} + \frac{3}{1 - 2v} \frac{\partial R}{\partial p_\alpha} \frac{\partial R}{\partial p_\eta} \right\} \right], \qquad (8.2.5)$$

$$g(\mathbf{p}_\alpha, \mathbf{q}_\eta)^* = \frac{(2v - 1)}{2(1 - v)} \frac{1}{R^2} \left[\frac{\partial R}{\partial q_\eta} n_\alpha - \frac{\partial R}{\partial q_\alpha} n_\eta \right.$$
$$\left. + \frac{\partial R}{\partial n} \left\{ \delta_{\alpha\eta} + \frac{3}{1 - 2v} \frac{\partial R}{\partial q_\eta} \frac{\partial R}{\partial q_\alpha} \right\} \right], \qquad (8.2.6)$$

and there is no difficulty in proving directly that the vector $g(\mathbf{p}_\alpha, \mathbf{q}_\eta)^*$; $\alpha = 1, 2, 3$ satisfies equation (8.1.4) for fixed $\mathbf{q}$ and η.

We now integrate $g^*(\mathbf{p}_\alpha, \mathbf{q}_\eta)$ over a large sphere ∂S of radius R centred upon $\mathbf{q} = 0$. In this case

$$\frac{\partial R}{\partial n} = -1, \qquad \frac{\partial R}{\partial p_\alpha} = \frac{p_\alpha}{R}, \qquad \frac{\partial R}{\partial p_\eta} = \frac{p_\eta}{R}, \qquad (8.2.7)$$

$$\int_{\partial S} \frac{\partial R}{\partial p_\alpha} dp = 0, \qquad \int_{\partial S} \frac{\partial R}{\partial p_\eta} dp = 0, \qquad \int_{\partial S} \frac{\partial R}{\partial p_\alpha} \frac{\partial R}{\partial p_\eta} dp = \frac{4\pi}{3} R^2 \delta_{\alpha\eta}, \qquad (8.2.8)$$

where the results (8.2.8) follow most simply from symmetry considerations. Accordingly

$$\int_{\partial S} g^*(\mathbf{p}_\alpha, \mathbf{q}_\eta) \, dp = -\frac{(2v-1)}{2(1-v)} \frac{1}{R^2} 4\pi R^2 \left(1 + \frac{1}{1-2v}\right) \delta_{\alpha\eta} = 4\pi \delta_{\alpha\eta}, \qquad (8.2.9)$$

as already anticipated.

The two-dimensional expressions corresponding to (8.2.3), etc., are

$$g(\mathbf{p}_\alpha \cdot \mathbf{q}_\eta) = -\frac{3-4v}{4(1-v)\mu} \delta_{\alpha\eta} \log R + \frac{1}{4(1-v)\mu} \frac{\partial R}{\partial p_\alpha} \frac{\partial R}{\partial p_\eta}, \qquad (8.2.10)$$

$$g^*(\mathbf{p}_\alpha, \mathbf{q}_\eta) = \frac{1-2v}{4(1-v)} \frac{1}{R} \left[\frac{\partial R}{\partial p_\alpha} n_\eta - \frac{\partial R}{\partial p_\eta} n_\alpha \right.$$

$$\left. + \frac{\partial R}{\partial n} \left\{ \delta_{\alpha\eta} + \frac{2}{1-2v} \frac{\partial R}{\partial p_\alpha} \frac{\partial R}{\partial p_\eta} \right\} \right], \qquad (8.2.11)$$

$$g(\mathbf{p}_\alpha, \mathbf{q}_\eta)^* = \frac{1-2v}{4(1-v)} \frac{1}{R} \left[\frac{\partial R}{\partial q_\eta} n_\alpha - \frac{\partial R}{\partial q_\alpha} n_\eta \right.$$

$$\left. + \frac{\partial R}{\partial n} \left\{ \delta_{\alpha\eta} + \frac{2}{1-2v} \frac{\partial R}{\partial q_\eta} \frac{\partial R}{\partial q_\alpha} \right\} \right]. \qquad (8.2.12)$$

These expressions refer to a concentrated force of strength 2π per unit height of an infinite cylinder of isotropic elastic material.

8.3 Biharmonic analysis

It follows either directly from equation (8.1.4), or from (8.1.7) and (8.1.8), that

$$\nabla^2(\nabla \cdot \boldsymbol{\phi}) = 0, \qquad \nabla^4 \boldsymbol{\phi} = 0, \qquad (8.3.1)$$

which shows that $\boldsymbol{\phi}$ is a specialised biharmonic vector. This suggests a study of the biharmonic scalar equation

$$\nabla^4 \chi \equiv \nabla^2 \nabla^2 \chi = 0. \tag{8.3.2}$$

A general solution of this equation has been given by Almansi (1897), viz,

$$\chi = x\phi + \psi; \qquad \nabla^2\phi = 0, \qquad \nabla^2\psi = 0, \tag{8.3.3}$$

or equivalently

$$\chi = y\phi + \psi, \qquad \chi = z\phi + \psi, \tag{8.3.4}$$

where ϕ, ψ are harmonic functions (we use x, y, z here in preference to p_1, p_2, p_3). To prove that (8.3.3) provides a solution for arbitrary ϕ and ψ, we note that

$$\nabla^2 \chi = 2 \frac{\partial \phi}{\partial x}, \tag{8.3.5}$$

$$\nabla^2(\nabla^2 \chi) = \nabla^2 \left(2 \frac{\partial \phi}{\partial x} \right) = 0. \tag{8.3.6}$$

To demonstrate its generality, we note that $\nabla^2 \chi$ is a harmonic function and that any harmonic function may be written as the derivative of another harmonic function (Bhattacharyya, 1975). Therefore equation (8.3.2) implies equation (8.3.5), which is a Poisson's equation for χ in terms of ϕ with a general solution given by (8.3.3).

The harmonic functions ϕ, ψ are not unique for a given χ, because the homogeneous equation

$$x\phi + \psi = 0; \qquad \nabla^2\phi = 0, \qquad \nabla^2\psi = 0, \tag{8.3.7}$$

is satisfied by

$$\phi = f(y, z), \qquad \psi = -xf(y, z); \qquad \nabla^2 f = 0, \tag{8.3.8}$$

where f is an arbitrary harmonic function of the variables y, z.

The representation

$$\chi = r^2\phi + \psi; \qquad \nabla^2\phi = 0, \qquad \nabla^2\psi = 0, \qquad r^2 = x^2 + y^2 + z^2, \tag{8.3.9}$$

is essentially equivalent to (8.3.3) or (8.3.4), but it has the advantage that ϕ, ψ are unique for a given χ. First, we note that

$$\nabla^2 \chi = \nabla^2(r^2 \phi) = 6\phi + 4\left(x\frac{\partial \phi}{\partial x} + y\frac{\partial \phi}{\partial y} + z\frac{\partial \phi}{\partial z}\right), \qquad (8.3.10)$$

$$\nabla^2(\nabla^2 \chi) = 6\nabla^2 \phi + 4\nabla^2\left(x\frac{\partial \phi}{\partial x} + y\frac{\partial \phi}{\partial y} + z\frac{\partial \phi}{\partial z}\right) = 0, \qquad (8.3.11)$$

which shows that (8.3.9) provides a biharmonic function for arbitrary ϕ and ψ. To demonstrate its generality, we prove that an arbitrary harmonic function may be expressed in the form (8.3.10). Thus, if χ (and hence also $\nabla^2 \chi$) is given within some domain, relation (8.3.10) may be viewed as a differential equation for ϕ in terms of $\nabla^2 \chi$ within this domain. To fix ideas, suppose that $\nabla^2 \chi$ is given within a sphere B. Then, following (8.1.20) and (8.1.21), we may write

$$\nabla^2 \chi = \sum_{n=0}^{\infty} \sum_{m=-n}^{n} b_n^m r^n S_n^m, \qquad \phi = \sum_{n=0}^{\infty} \sum_{m=-n}^{n} a_n^m r^n S_n^m \qquad (8.3.12)$$

within B, where the a_n^m are coefficients to be determined in terms of the b_n^m. Substituting (8.3.12) into (8.3.10) and equating coefficients of $r^n S_n^m$ on each side, we find

$$b_n^m = (6 + 4n)a_n^m, \quad \text{i.e. } a_n^m = b_n^m/(6 + 4n); \qquad n = 0, 1, 2, \ldots \qquad (8.3.13)$$

Accordingly the second expansion (8.3.12) certainly converges and therefore defines a unique harmonic function within B. An immediate deduction from (8.3.13) is that $\phi = 0$ if $\nabla^2 \chi = 0$, showing that the homogeneous equation

$$r^2 \phi + \psi = 0; \qquad \nabla^2 \phi = 0, \qquad \nabla^2 \psi = 0, \qquad (8.3.14)$$

has no solution in B other than $\phi = 0$, $\psi = 0$.

If B is the infinite domain exterior to a sphere, the expansions

$$\nabla^2 \chi = r^{-1} \sum_{n=0}^{-\infty} \sum_{m=-n}^{n} b_n^m r^n S_n^m, \qquad \phi = r^{-1} \sum_{n=0}^{-\infty} \sum_{m=-n}^{n} a_n^m r^n S_n^m \qquad (8.3.15)$$

replace those of (8.3.12). A straightforward repetition of the previous analysis then yields

$$b_n^m = \{6 + 4(n - 1)\}a_n^m, \quad \text{i.e. } a_n^m = b_n^m/(2 + 4n)$$

$$n = 0, -1, -2, \ldots \qquad\qquad\qquad\qquad\qquad\qquad (8.3.16)$$

which again provides a convergent expansion for ϕ within B. These conclusions are generalised in Appendix 8 to a domain of any shape.

The function

$$r^2 r^{-1} = r = |\mathbf{p}| \qquad\qquad (8.3.17)$$

is biharmonic everywhere except at $r = 0$, and it satisfies the differential equations

$$\nabla^2 r = 2r^{-1}, \qquad \nabla^4 r = 2\nabla^2(r^{-1}) = -8\pi\delta(\mathbf{p}). \qquad (8.3.18)$$

More generally, the function

$$R = |\mathbf{p} - \mathbf{q}| \qquad\qquad (8.3.19)$$

satisfies the differential equations

$$\nabla^2 R = 2|\mathbf{p} - \mathbf{q}|^{-1}, \qquad \nabla^4 R = -8\pi\delta(\mathbf{p} - \mathbf{q}) \qquad (8.3.20)$$

everywhere. By analogy with harmonic potential theory, we may regard (8.3.19) as the potential generated by a unit biharmonic source at $\mathbf{q}$. Pursuing this analogy, we introduce the simple-layer biharmonic potential

$$\Omega(\mathbf{p}) = \int_{\partial B} |\mathbf{p} - \mathbf{q}|\sigma(\mathbf{q})\,dq; \qquad \mathbf{p} \in B + \partial B, \qquad (8.3.21)$$

which has the property

$$\nabla^2\Omega(\mathbf{p}) = \int_{\partial B} \nabla^2|\mathbf{p} - \mathbf{q}|\sigma(\mathbf{q})\,dq = 2\int_{\partial B} |\mathbf{p} - \mathbf{q}|^{-1}\sigma(\mathbf{q})\,dq; \quad \mathbf{p} \in B. \quad (8.3.22)$$

Bearing in mind that $\nabla^2\chi$ can always be represented as a simple-layer potential in B, where χ is an arbitrary biharmonic function, it follows from (8.3.22) that we may write

$$\chi = \Omega + \psi; \qquad \nabla^2 \psi = 0. \tag{8.3.23}$$

This provides a valid alternative representation to (8.3.9).

8.4 Two-dimensional biharmonic analysis

The two-dimensional representation

$$\chi = x\phi + \psi; \qquad \nabla^2 \phi = 0, \qquad \nabla^2 \psi = 0, \tag{8.4.1}$$

has much the same features as (8.3.3), except that the non-uniqueness possibility (8.3.8) gets replaced by

$$\phi = a + by, \qquad \psi = -(ax + bxy), \tag{8.4.2}$$

where a, b are arbitrary constants. Similar considerations hold for

$$\chi = y\phi + \psi; \qquad \nabla^2 \phi = 0, \qquad \nabla^2 \psi = 0. \tag{8.4.3}$$

However the analysis of

$$\chi = r^2 \phi + \psi; \qquad \nabla^2 \phi = 0, \qquad \nabla^2 \psi = 0, \qquad r^2 = x^2 + y^2, \tag{8.4.4}$$

differs somewhat from that of (8.3.9), essentially because equation (8.3.10) gets replaced by

$$\nabla^2 \chi = 4\phi + 4\left(x \frac{\partial \phi}{\partial x} + y \frac{\partial \phi}{\partial y} \right). \tag{8.4.5}$$

To solve this equation for ϕ within a circle B, we introduce the convergent expansion

$$\nabla^2 \chi = \sum_{n=0}^{\infty} r^n (b_{nc} \cos n\theta + b_{ns} \sin n\theta) \tag{8.4.6}$$

within B, where b_{nc}, b_{ns} are uniquely determined coefficients. Also we assume the corresponding expansion

$$\phi = \sum_{n=0}^{\infty} r^n(a_{nc} \cos n\theta + a_{ns} \sin n\theta) \qquad (8.4.7)$$

within B, where a_{nc}, a_{ns} are coefficients to be determined in terms of b_{nc}, b_{ns}. Substituting (8.4.6), (8.4.7) into (8.4.5), and equating coefficients of $r_n \cos n\theta$, $r^n \sin n\theta$ on each side, we find

$$\left.\begin{array}{ll} b_{nc} = (4 + 4n)a_{nc}, & \text{i.e. } a_{nc} = b_{nc}/(4 + 4n) \\[2mm] b_{ns} = (4 + 4n)a_{ns}, & \text{i.e. } a_{ns} = b_{ns}/(4 + 4n) \end{array}\right\} n = 0, 1, 2, \ldots. \qquad (8.4.8)$$

Accordingly, the expansion (8.4.7) certainly converges and therefore defines a unique harmonic function ϕ within B.

If B is the infinite domain exterior to a circle, the relevant expansions are

$$\left.\begin{array}{l} \nabla^2 \chi = \sum_{n=0}^{-\infty} r^n(b_{nc} \cos n\theta + b_{ns} \sin n\theta) \\[4mm] \phi = \sum_{n=0}^{-\infty} r^n(a_{nc} \cos n\theta + a_{ns} \sin n\theta) \end{array}\right\}. \qquad (8.4.9)$$

These yield equations formally similar to (8.4.8) except that now

$$n = 0, -1, -2, \ldots, \qquad (8.4.10)$$

which shows that a breakdown occurs when $n = -1$, i.e. when $\nabla^2 \chi$ contains components proportional to $r^{-1} \cos \theta$, $r^{-1} \sin \theta$. Since

$$\nabla^2(x \log r) = 2r^{-1} \cos \theta, \qquad \nabla^2(y \log r) = 2r^{-1} \sin \theta, \qquad (8.4.11)$$

it follows that the singular biharmonic functions $x \log r$, $y \log r$ cannot necessarily be represented by (8.4.4). These functions are inadmissible within any domain enclosing $r = 0$, but they could exist within any domain which excludes $r = 0$.

It may be noted that

$$
\left.
\begin{aligned}
2x \log r &= r^2 \left(\frac{x \log r + y\theta}{r^2} \right) + (x \log r - y\theta) \\[2ex]
2y \log r &= r^2 \left(\frac{y \log r - x\theta}{r^2} \right) + (y \log r + x\theta)
\end{aligned}
\right\}, \qquad (8.4.12)
$$

where

$$
\phi = \frac{x \log r + y\theta}{r^2}, \qquad \frac{y \log r - x\theta}{r^2} \qquad (8.4.13)
$$

$$
\psi = x \log r - y\theta, \qquad y \log r + x\theta \qquad (8.4.14)
$$

are multi-valued harmonic functions for any circuit around the origin. Accordingly $x \log r$, $y \log r$ can be represented in the form (8.4.4) provided we admit multi-valued harmonic functions. Operating upon both sides of (8.4.12) by ∇^2, we infer that ϕ as given by (8.4.13) satisfies the equation

$$
4r^{-1} \cos \theta, \qquad 4r^{-1} \sin \theta = 4\phi + 4\left(x \frac{\partial \phi}{\partial x} + y \frac{\partial \phi}{\partial y} \right), \qquad (8.4.15)
$$

and this may be verified directly.

By contrast with (8.3.14), the two-dimensional homogeneous equation

$$
r^2 \phi + \psi = 0; \qquad \nabla^2 \phi = 0, \qquad \nabla^2 \psi - 0, \qquad (8.4.16)
$$

has the independent non-trivial solutions

$$
\left.
\begin{aligned}
\phi &= r^{-1} \cos \theta, \qquad \psi = -r \cos \theta = -x \\[2ex]
\phi &= r^{-1} \sin \theta, \qquad \psi = -r \sin \theta = -y
\end{aligned}
\right\}.
\qquad (8.4.17)
$$

Operating upon both sides of (8.4.16) by ∇^2, we infer that ϕ as given by (8.4.17) satisfies the equation

$$
4\phi + 4\left(x \frac{\partial \phi}{\partial x} + y \frac{\partial \phi}{\partial y} \right) = 0. \qquad (8.4.18)
$$

This may be verified directly and also by putting b_{nc}, $b_{ns} = 0$ in (8.4.8).

The two-dimensional function $r^2 \log r$ corresponds to (8.3.17). More generally, the function

$$R^2 \log R = |\mathbf{p} - \mathbf{q}|^2 \log|\mathbf{p} - \mathbf{q}| \qquad (8.4.19)$$

corresponds to (8.3.19). This satisfies the differential equations

$$\nabla^2(R^2 \log R) = 4 + 4 \log R, \qquad (8.4.20)$$

$$\nabla^4(R^2 \log R) = 4\nabla^2 \log R = 8\pi\delta(\mathbf{p} - \mathbf{q}). \qquad (8.4.21)$$

Accordingly we may regard (8.4.19) as the two-dimensional potential generated by a unit biharmonic source at $\mathbf{q}$. It has been suggested by Chakrabarty (1971) that a preferable potential is

$$G(\mathbf{p}, \mathbf{q}) = -R^2 + R^2 \log R, \qquad (8.4.22)$$

since this satisfies the equations

$$\nabla^2 G = 4 \log R, \qquad \nabla^4 G = 8\pi\delta(\mathbf{p} - \mathbf{q}). \qquad (8.4.23)$$

Corresponding to (8.3.21) we therefore introduce the simple-layer biharmonic potential

$$\Omega(\mathbf{p}) = \int_{\partial B} G(\mathbf{p}, \mathbf{q})\sigma(\mathbf{q}) \, dq; \qquad \mathbf{p} \in B_i + \partial B, \qquad (8.4.24)$$

which has the convenient property

$$\nabla^2\Omega(\mathbf{p}) = \int_{\partial B} \nabla^2 G(\mathbf{p}, \mathbf{q})\sigma(\mathbf{q}) \, dq = 4 \int_{\partial B} \log|\mathbf{p} - \mathbf{q}|\sigma(\mathbf{q}) \, dq; \qquad \mathbf{p} \in B_i. \quad (8.4.25)$$

Biharmonic potentials may exist within the infinite domain B_e exterior to ∂B, and

$$\Omega(\mathbf{p}) = (-r^2 + r^2 \log r) \int_{\partial B} \sigma(\mathbf{q}) \, dq - 2x \log r \int_{\partial B} q_1 \sigma(\mathbf{q}) \, dq$$

$$- 2y \log r \int_{\partial B} q_2 \sigma(\mathbf{q}) \, dq + \frac{r^2}{4} \sum_{n=-2}^{-\infty} \frac{r^n}{n+1} (b_{nc} \cos n\theta + b_{ns} \sin n\theta)$$

$$+ \psi_1 \quad \text{as} \quad |\mathbf{p}| = r \to \infty, \tag{8.4.26}$$

where q_1, q_2 are the Cartesian coordinates of $\mathbf{q}$ and ψ_1 is a harmonic function of the form $\alpha x + \beta y + \gamma + O(r^{-1})$. This may be demonstrated directly by examining the asymptotic behaviour of $G(\mathbf{p}, \mathbf{q})$, and it also follows by noting from (8.4.25) and (4.1.3) that

$$\nabla^2 \Omega(\mathbf{p}) = 4 \log r \int_{\partial B} \sigma(\mathbf{q}) \, dq - 4 \frac{\cos \theta}{r} \int_{\partial B} q_1 \sigma(\mathbf{q}) \, dq - 4 \frac{\sin \theta}{r} \int_{\partial B} q_2 \sigma(\mathbf{q}) \, dq$$

$$+ \sum_{n=-2}^{-\infty} r^n (b_{nc} \cos n\theta + b_{ns} \sin n\theta) \quad \text{as} \quad |\mathbf{p}| = r \to \infty. \tag{8.4.27}$$

Within a ring-shaped domain B bounded externally by ∂B_0 and internally ∂B_1 which encloses $r = 0$, we write

$$\Omega(\mathbf{p}) = \int_{\partial B_0} G(\mathbf{p}, \mathbf{q}) \sigma(\mathbf{q}) \, dq + \int_{\partial B_1} G(\mathbf{p}, \mathbf{q}) \sigma(\mathbf{q}) \, dq; \quad \mathbf{p} \in B + \partial B_0 + \partial B_1.$$

$$\tag{8.4.28}$$

The first of these biharmonic potentials has the form

$$\int_{\partial B_0} G(\mathbf{p}, \mathbf{q}) \sigma(\mathbf{q}) \, dq = \frac{r^2}{4} \sum_{n=0}^{\infty} \frac{r^n}{n+1} (b_{nc} \cos n\theta + b_{ns} \sin n\theta) + \psi_0;$$

$$\mathbf{p} \in B, \tag{8.4.29}$$

where ψ_0 is a harmonic function of the form (8.4.7) in B, and the second has the form (8.4.26).

9

Isotropic Elasticity: Two-Dimensional Theory

9.1 Two-dimensional elastic systems

The three-dimensional theory in principle covers all two-dimensional boundary-value problems, but an alternative approach to the latter is through the biharmonic scalar functions introduced in Section 8.4. Linear problems of the bending and stretching of plates are governed by equation (8.3.2), which holds throughout the domain B of the plate subject to suitable boundary conditions along the edge ∂B. Interesting differences emerge between the bending and stretching problems in the case of a ring-shaped domain.

Hadamard (1908) proved that a function χ, which is continuous and has continuous derivatives up to the fourth order within an interior domain B_i, and which satisfies equation (8.3.2) in B_i, is uniquely determined in B_i if it has a prescribed continuous set of values χ and of normal derivative values χ' (subscript i understood) on ∂B. To construct χ in B_i we utilise the Almansi representation (8.4.4), taking the origin $r = 0$ inside B_i. This yields the boundary functional relations

$$\chi = r^2\phi + \psi, \tag{9.1.1}$$

$$\chi' = (r^2\phi)' + \psi' = 2rr'\phi + r^2\phi' + \psi' \tag{9.1.2}$$

connecting ϕ, ϕ', ψ, ψ' on ∂B. Also ϕ, ϕ' are connected by the two-dimensional Green's boundary formula (4.4.3), which we symbolise

$$[\phi, \phi'] - \pi\phi = 0, \tag{9.1.3}$$

and similarly

$$[\psi, \psi'] - \pi\psi = 0. \tag{9.1.4}$$

From (9.1.1), (9.1.2) and (9.1.4) we obtain

$$[\chi, \chi'] - \pi\chi = [r^2\phi, (r^2\phi)'] - \pi r^2\phi \tag{9.1.5}$$

which, together with (9.1.3), provides a pair of coupled boundary integral equations for ϕ, ϕ'. With these known, we determine ψ, ψ' from (9.1.1), (9.1.2) whence Green's formula (4.4.1) enables us to generate ϕ and ψ, and hence also χ, throughout B_i (Jaswon, 1963).

An alternative method for constructing χ is to write

$$\phi(\mathbf{p}) = \int_{\partial B} \log|\mathbf{p} - \mathbf{q}|\sigma(\mathbf{q})\,dq; \qquad \mathbf{p} \in B_i + \partial B, \tag{9.1.6}$$

$$\psi(\mathbf{p}) = \int_{\partial B} \log|\mathbf{p} - \mathbf{q}|\zeta(\mathbf{q})\,dq; \qquad \mathbf{p} \in B_i + \partial B, \tag{9.1.7}$$

where σ, ζ are hypothetical source densities to be determined. We bear in mind the normal derivative expressions

$$\phi'(\mathbf{p}) = \int_{\partial B} \log'|\mathbf{p} - \mathbf{q}|\sigma(\mathbf{q})\,dq + \pi\sigma(\mathbf{p}); \qquad \mathbf{p} \in \partial B, \tag{9.1.8}$$

$$\psi'(\mathbf{p}) = \int_{\partial B} \log'|\mathbf{p} - \mathbf{q}|\zeta(\mathbf{q})\,dq + \pi\zeta(\mathbf{p}); \qquad \mathbf{p} \in \partial B. \tag{9.1.9}$$

Substituting from (9.1.6)–(9.1.9) into (9.1.1), (9.1.2) yields two coupled integral equations for σ, ζ which may be solved numerically (see Chapter 15) so enabling ϕ and ψ, and hence also χ, to be generated numerically throughout B_i. A specialised problem is provided by the boundary data

$$\chi = \alpha x + \beta y + \gamma, \qquad \chi' = \alpha x' + \beta y'; \qquad (x, y) \in \partial B, \tag{9.1.10}$$

where α, β, γ are constants. Clearly

$$\chi = \alpha x + \beta y + \gamma; \qquad (x, y) \in B_i + \partial B, \tag{9.1.11}$$

E

with the breakdown

$$\phi = 0, \qquad \psi = \alpha x + \beta y + \gamma; \qquad (x, y) \in B_i + \partial B. \qquad (9.1.12)$$

The uniqueness of χ for an infinite exterior domain B_e requires

$$\chi = O(r) \quad \text{as} \quad r \to \infty, \qquad (9.1.13)$$

in addition to the previously mentioned requirements for B_i. This covers the Stokes' paradox of slow viscous flow (Lamb, 1945) where χ stands for the stream function with velocity components given by $-\partial\chi/\partial y, \partial\chi/\partial x$. If the flow is perturbed by a rigid internal boundary ∂B then we may put $\chi = 0$, $\partial\chi/\partial n = 0$ along ∂B. These conditions and condition (9.1.13) are satisfied by $\chi = 0$ everywhere in $B_e + \partial B$, which must therefore be the only possible solution. This conclusion may also be interpreted in terms of the transverse deflection of a thin plate clamped at an internal boundary (Section 9.2). Requirement (9.1.13) is ensured by the representation (8.4.4) provided that $\phi = O(r^{-1})$ as $r \to \infty$. We therefore utilise (9.1.6) subject to the condition

$$\int_{\partial B} \sigma(\mathbf{q}) \, dq = 0, \qquad (9.1.14)$$

since then

$$\phi = O(r^{-1}), \qquad r^2\phi = O(r) \quad \text{as} \quad r \to \infty. \qquad (9.1.15)$$

Also it is advantageous to utilise (9.1.7) since this

(a) yields the acceptable behaviour $\psi = O(\log r)$ as $r \to \infty$ and
(b) excludes the non-uniqueness possibilities (8.4.17) which imply $\psi = O(r)$ as $r \to \infty$.

However, unlike the interior problem, ψ does not now cover the possibility of a constant component entering into χ. We therefore write

$$\chi = r^2\phi + \psi + k \quad \text{in} \quad B_e + \partial B, \qquad (9.1.16)$$

where k is a constant to be determined. This extra unknown balances the side condition (9.1.14) but otherwise the formulation proceeds on much the same lines as previously. A useful specialised problem is again provided by (9.1.10), though of course the normal direction now points into B_e. Clearly

$$\chi = \alpha x + \beta y + \gamma; \qquad (x, y) \in B_e + \partial B, \tag{9.1.17}$$

with the breakdown (utilising polar coordinates)

$$\phi = \alpha \frac{\cos \theta}{r} + \beta \frac{\sin \theta}{r}, \qquad \psi = 0, \qquad k = \gamma, \tag{9.1.18}$$

since ψ cannot cover $\alpha x + \beta y$ in B_e.

For a ring-shaped domain D bounded externally by ∂B_0 and internally by ∂B_1 which encloses $r = 0$, the special biharmonic functions $x \log r$, $y \log r$ cannot be covered by (8.4.4) if we utilise the inherently single-valued harmonic functions (9.1.6), (9.1.7). Accordingly we write

$$\chi = r^2 \phi + \psi + ax \log r + by \log r; \qquad (x, y) \in B + \partial B, \tag{9.1.19}$$

where ϕ, ψ are defined by (9.1.6), (9.1.7) on the understanding that $\partial B = \partial B_0 + \partial B_1$. The unknown coefficients a, b may be balanced by the introduction of suitable side conditions. These must be chosen so as to eliminate the nonuniqueness possibilities (8.4.17), i.e. to ensure that ϕ does not contain any components proportional to $r^{-1} \cos \theta$, $r^{-1} \sin \theta$. Reference to (4.1.3) shows that these conditions are

$$\int_{\partial B_1} q_1 \sigma(\mathbf{q}) \, dq = 0, \qquad \int_{\partial B_1} q_2 \sigma(\mathbf{q}) \, dq = 0; \qquad \mathbf{q} = (q_1, q_2), \tag{9.1.20}$$

it being noted that asymptotic behaviour depends only on the source distribution over ∂B_1. Coupling (9.1.20) with the generalised versions of (9.1.1), (9.1.2) corresponding to (9.1.19), we have sufficient equations to determine a, b and σ, ζ, whence ϕ and ψ, and hence also χ, may be generated throughout B.

Biharmonic boundary-value problems may also be formulated by utilising the Chakrabarty representation (8.4.24), i.e. we write

$$\chi = \Omega + \psi; \qquad \nabla^2 \psi = 0, \quad \text{in} \quad B_i + \partial B. \tag{9.1.21}$$

The boundary equations then become

$$\chi = \Omega + \psi, \qquad \chi' = \Omega' + \psi' \quad \text{on} \quad \partial B \tag{9.1.22}$$

in place of (9.1.1), (9.1.2). An analysis of Ω, Ω' as $\mathbf{p} \to \mathbf{q} \in \partial B$ is available (Bhattacharyya, 1975). For exterior problems, following (8.4.26),

$$\chi = \Omega + \psi + \alpha x + \beta y + \gamma; \qquad (x, y) \in B_e + \partial B, \qquad (9.1.23)$$

where α, β, γ are constants to be determined subject to the side conditions

$$\int_{\partial B} \sigma(\mathbf{q}) \, dq = 0, \qquad \int_{\partial B} q_1 \sigma(\mathbf{q}) \, dq = 0, \qquad \int_{\partial B} q_2 \sigma(\mathbf{q}) \, dq = 0 \quad (9.1.24)$$

which ensure $\chi = O(r)$ as $r \to \infty$. Within a ring-shaped domain, Ω as defined by (8.4.28) covers $r^2 \log r$, $x \log r$, $y \log r$ and therefore (9.1.21) applies on the understanding that $\partial B = \partial B_0 + \partial B_1$. Accordingly, the Chakrabarty representation is seen to be superior to that of Almansi for a ring-shaped domain but inferior for an infinite exterior domain. Apart from these two representations, the Rayleigh-Green identity provides an interesting alternative approach to two-dimensional boundary-value problems (see Appendix 9).

9.2 Bending of thin plates

The small transverse deflection, w, of a thin isotropic elastic plate under the action of a transverse load T per unit area, satisfies the equation

$$\nabla^4 w = T/D \quad \text{in} \quad B_i \text{ (the domain of the plate),} \qquad (9.2.1)$$

where D is the flexural rigidity ($D = 2h^3 E/3(1 - v^2)$, E is Young's modulus, v is Poisson's ratio, and $2h$ is the plate thickness). This equation has a general solution of the form

$$w = W + \chi, \qquad (9.2.2)$$

where W is a particular solution and χ satisfies equation (8.3.2). For a uniform transverse load, i.e. $T = k$ (constant), we have

$$W(\mathbf{p}) = \frac{k}{64D}(x^2 + y^2)^2 \qquad \text{or} \quad W(\mathbf{p}) = \frac{k}{48D}(x^4 + y^4) \qquad (9.2.3)$$

at $\mathbf{p} = (x, y)$. For a concentrated transverse load applied at some point

$\mathbf{q} \in B_i$, i.e. expressed mathematically

$$T(\mathbf{p}) = k\delta(\mathbf{p} - \mathbf{q}); \qquad k = \text{total load,} \qquad (9.2.4)$$

we have

$$W(\mathbf{p}) = \frac{k}{8\pi D}|\mathbf{p} - \mathbf{q}|^2 \log|\mathbf{p} - \mathbf{q}| \qquad (9.2.5)$$

bearing in mind (8.4.21).

The boundary conditions for a clamped plate are

$$w = 0, \qquad w' = 0 \qquad \text{on} \qquad \partial B, \qquad (9.2.6)$$

i.e. $\qquad\qquad W + \chi = 0, \qquad W' + \chi' = 0 \qquad \text{on} \qquad \partial B,$

or $\qquad\qquad \chi = -W, \qquad \chi' = -W' \qquad \text{on} \qquad \partial B, \qquad (9.2.7)$

Since W, W' may be computed on ∂B, it follows that χ, χ' are known on ∂B, so enabling χ to be determined throughout B_i by any of the methods of Section 9.1 whence w may be generated throughout B_i from (9.2.2). Utilising an obvious notation, the associated bending and twisting moments on ∂B appear, respectively, as

$$M_{nn} = -D\left(\nabla^2 w + \frac{1 - v}{\rho}\frac{\partial w}{\partial n}\right) \qquad (9.2.8)$$

$$M_{nt} = -D(1 - v)\frac{\partial^2 w}{\partial n\partial t}$$

where ρ is the internal radius of curvature. These expressions simplify to

$$M_{nn} = -D\nabla^2 w, \qquad M_{nt} = 0 \qquad (9.2.9)$$

on bearing in mind that $\partial w/\partial n = 0$, $\partial w/\partial t = 0$ under clamped conditions.

The boundary conditions for a simply-supported plate are

$$w = 0, \qquad M_{nn} = 0. \qquad (9.2.10)$$

Substituting $w = W + \chi$ into (9.2.10), we formulate two relations between χ, χ' on ∂B which suffice to determine σ, ζ and hence eventually w throughout B_i. The interesting boundary quantities $\partial w/\partial n$, M_{nt} may then be calculated. It will be noticed that $\rho^{-1} \partial w/\partial n = 0$ along a straight line boundary, in which case conditions (9.2.10) reduce to

$$w = 0, \qquad \nabla^2 w = 0. \qquad (9.2.11)$$

This greatly simplifies the problem for polygonal boundaries, on the assumption that corner effects may be neglected.

The theory of freely-supported plates offers considerable analytical and numerical difficulties, which have not yet been completely surmounted (see E. Jaswon, 1973).

9.3 Stretching of thin plates: traction problems

Here χ represents the Airy stress function. This has no direct physical or geometrical significance, but its derivatives provide the stress components

$$\Phi_{xx} = \frac{\partial^2 \chi}{\partial y^2}, \qquad \Phi_{yy} = \frac{\partial^2 \chi}{\partial x^2}, \qquad \Phi_{xy} = -\frac{\partial^2 x}{\partial x \partial y}. \qquad (9.3.1)$$

These readily yield the boundary quantities

$$\frac{\partial \chi}{\partial x} = -\int_\pi^{\mathbf{p}} T_y(\mathbf{q}) \, dq + \alpha, \qquad \frac{\partial \chi}{\partial y} = \int_\pi^{\mathbf{p}} T_x(\mathbf{q}) \, dq + \beta \qquad (9.3.2)$$

where T_x, T_y are the given (mechanical) traction components per unit length of arc, π signifies an arbitrary origin of integration on ∂B, $\mathbf{p}$ signifies the current point on ∂B at which $\partial \chi/\partial x$, $\partial \chi/\partial y$ are being evaluated, and α, β are arbitrary constants of integration. An integral equation formulation in which T_x, T_y appear directly has been given by Weinel (1931). We thus arrive at the canonical boundary quantities

$$\chi' \equiv \frac{\partial \chi}{\partial n} = \frac{\partial \chi}{\partial x} \frac{\partial x}{\partial n} + \frac{\partial \chi}{\partial y} \frac{\partial y}{\partial n}, \qquad (9.3.3)$$

$$\chi = \int_\pi^\mathbf{p} \frac{\partial \chi}{\partial q} \, dq + \gamma = \int_\pi^\mathbf{p} \left(\frac{\partial \chi}{\partial x} \frac{\partial x}{\partial q} + \frac{\partial \chi}{\partial y} \frac{\partial y}{\partial q} \right) dq + \gamma, \qquad (9.3.4)$$

where γ is a further constant of integration. The expression for $\partial\chi/\partial n$ includes a term $\alpha(\partial x/\partial n) + \beta(\partial y/\partial n)$ arising from α, β and that for χ includes a corresponding term $\alpha x + \beta y + \gamma$. Reference to (9.1.11) shows that these boundary contributions define a trivial biharmonic function in B_i which yields no stress components. Accordingly, neither the location of π nor the values there of $\partial\chi/\partial x, \partial\chi/\partial y, \chi$ have any physical significance.

It will be seen that $[\partial\chi/\partial y]_\pi^\mathbf{p}$ measures the x-component of the resultant force acting upon ∂B between π and $\mathbf{p}$. If, therefore, the total x-component vanishes, as must be the case for a simply-connected domain in statical equilibrium, then $\partial\chi/\partial y$ is a single-valued function on ∂B. Similar considerations hold for $\partial\chi/\partial x$. Also the resultant couple acting upon ∂B between π and $\mathbf{p}$ is

$$\int_\pi^\mathbf{p} (xT_y - yT_x) \, dq = - \int_\pi^\mathbf{p} \left\{ x \frac{\partial}{\partial q}\left(\frac{\partial \chi}{\partial x} \right) + y \frac{\partial}{\partial q}\left(\frac{\partial \chi}{\partial y} \right) \right\} dq$$

$$= \left[\chi - x \frac{\partial \chi}{\partial x} - y \frac{\partial \chi}{\partial y} \right]_\pi^\mathbf{p}. \qquad (9.3.5)$$

Accordingly, if the total resultant couple acting upon ∂B also vanishes, as must be the case for statical equilibrium, then χ is a single-valued function on ∂B and therefore also throughout B_i.

For exterior problems, the representation (9.1.16) or (9.1.23) applies coupled with the side condition (9.1.14) or (9.1.24) respectively.

For a ring-shaped domain B the possibility arises of a non-vanishing resultant force or couple acting upon ∂B_1 with a compensating force or couple acting upon ∂B_0, thus implying a multi-valued χ in B (see Section 9.4). Discounting this possibility, which will be considered later, we now extend the preceding boundary analysis as follows. Starting from an arbitrary origin π_0 on ∂B_0, we calculate $\partial\chi/\partial x, \partial\chi/\partial y, \chi$ as before. These quantities resume their starting values after a circuit of ∂B_0. We then jump from π_0 to some arbitrary origin π_1 on ∂B_1, it being supposed that $\partial\chi/\partial x, \partial\chi/\partial y, \chi$ jump to the values

$$\frac{\partial \chi}{\partial x} = \alpha, \qquad \frac{\partial \chi}{\partial y} = \beta, \qquad \chi = \gamma \qquad \text{at } \pi_1 = (x_1, y_1) \in \partial B_1,$$

thereby making contributions

$$\frac{\partial \chi}{\partial n} = \alpha \frac{\partial x}{\partial n} + \beta \frac{\partial y}{\partial n}, \qquad \chi = \alpha(x - x_1) + \beta(y - y_1) + \gamma \qquad (9.3.6)$$

to the canonical boundary quantities on ∂B_1. If ∂B_0, ∂B_1 are traction-free, the boundary data on ∂B_0 are $\chi = 0$, $\partial \chi / \partial n = 0$ which, together with (9.3.6), define a unique non-trivial χ in B, so implying the existence of a continuous elastic stress field generated entirely by internal sources. These sources are the three independent two-dimensional Volterra dislocations (Love, 1927), each characterised by a discontinuity in a specific displacement or rotation component. Analysis (next section) shows that such discontinuities are inherent features of the three functions

$$x \log r, \qquad y \log r, \qquad r^2 \log r \qquad (9.3.7)$$

which we therefore identify as dislocation solutions of the biharmonic equation. To eliminate unwanted dislocations, we put $a = b = 0$ in (9.1.19) and utilise the side conditions

$$\int_{\partial B_1} q_1 \sigma(\mathbf{q}) \, dq = 0, \qquad \int_{\partial B_1} q_2 \sigma(\mathbf{q}) \, dq = 0, \qquad \int_{\partial B_1} \sigma(\mathbf{q}) \, dq = 0 \qquad (9.3.8)$$

to balance the three boundary unknowns α, β, γ. These side conditions should be compared with those given by Southwell (1956). The third condition of (9.3.8) ensures that $r^2 \phi$ contains no component proportional to $r^2 \log r$, whilst the first two conditions ensure that ϕ, ψ are unique, cf. (9.1.20). A similar analysis holds for the representation (9.1.21), and indeed conditions (9.3.8) now assume a more direct significance in eliminating dislocations as follows from the asymptotic behaviour (8.4.26).

9.4 Stretching of thin plates: displacement problems

The displacement components u, v corresponding to χ may be expressed by

$$2\mu u = (1 - v) H - \frac{\partial \chi}{\partial x}, \qquad 2\mu v = (1 - v) \bar{H} - \frac{\partial \chi}{\partial x} \qquad (9.4.1)$$

where $H, \bar{H}$ are conjugate harmonic functions satisfying

$$\frac{\partial H}{\partial x} = \frac{\partial \bar{H}}{\partial y} = \nabla^2 \chi. \tag{9.4.2}$$

Here μ, v denote the shear modulus and Poisson's ratio respectively. A convenient representation for χ in this context is $\chi = x\phi + \psi$, since then $\nabla^2 \chi = 2(\partial\phi/\partial x)$ which implies $H = 2\phi$, $\bar{H} = 2\bar{\phi}$. Accordingly formulae (9.4.1) become

$$2\mu u = 2(1 - v)\phi - \frac{\partial\chi}{\partial x}, \qquad 2\mu v = 2(1 - v)\bar{\phi} - \frac{\partial\chi}{\partial y} \tag{9.4.3}$$

to which we add

$$2\mu\omega = -\mu\left(\frac{\partial u}{\partial y} - \frac{\partial v}{\partial x}\right) = -2(1 - v)\frac{\partial\phi}{\partial y} \tag{9.4.4}$$

for the local (anti-clockwise) rotation ω. A single-valued χ implies that $\partial\chi/\partial x$, $\partial\chi/\partial y$ are single-valued, but ϕ, $\bar{\phi}$, or $\partial\phi/\partial y$ may be multi-valued, the three independent possibilities being

$$
\begin{aligned}
&\chi = x\log r, &&\phi - \log r, &&\bar{\phi} = \theta, &&\psi - 0 \\
&\chi = -y\log r, &&\phi = 0, &&\bar{\phi} = -\log r, &&\psi = -(y\log r + x\theta) \\
&\chi = \tfrac{1}{2}r^2\log r, &&\phi = x\log r - y\theta, &&\bar{\phi} = y\log r + x\theta, \\
&&&&&&\psi = \tfrac{1}{2}(y^2 - x^2)\log r + xy\theta
\end{aligned}
\Biggr\}. \tag{9.4.5}
$$

These provide jumps of amount $2(1 - v)\pi/\mu$ in v, u, ω respectively on making an anti-clockwise circuit around the origin, so confirming the three dislocation solutions already introduced.

Given $2\mu u$, $2\mu v$ over ∂B, relations (9.4.3) constitute a pair of functional equations coupling ϕ, $\bar{\phi}$, $\partial\chi/\partial x$, $\partial\chi/\partial y$ on ∂B. Progress towards a solution of these equations may be achieved by writing

$$\frac{\partial\chi}{\partial x} = x\frac{\partial\phi}{\partial x} + \phi + \frac{\partial\psi}{\partial x}, \qquad \frac{\partial\chi}{\partial y} = x\frac{\partial\phi}{\partial y} + \frac{\partial\psi}{\partial y} \tag{9.4.6}$$

and noting that $\partial\psi/\partial x$, $-\partial\psi/\partial y$ are conjugate harmonic functions. This allows us to introduce the representations

$$\phi(\mathbf{p}) = \int_{\partial B} \log|\mathbf{p} - \mathbf{q}|\,\sigma(\mathbf{q})\,dq; \qquad \mathbf{p} \in B + \partial B, \tag{9.4.7}$$

$$\bar{\phi}(\mathbf{p}) = \int_{\partial B} \theta(\mathbf{p} - \mathbf{q})\,\sigma(\mathbf{q})\,dq; \qquad \mathbf{p} \in B + \partial B, \tag{9.4.8}$$

$$\frac{\partial\psi}{\partial x}(\mathbf{p}) = \int_{\partial B} \log|\mathbf{p} - \mathbf{q}|\,\zeta(\mathbf{q})\,dq; \qquad \mathbf{p} \in B + \partial B, \tag{9.4.9}$$

$$\frac{\partial\psi}{\partial y}(\mathbf{p}) = -\int_{\partial B} \theta(\mathbf{p} - \mathbf{q})\,\zeta(\mathbf{q})\,dq; \qquad \mathbf{p} \in B + \partial B, \tag{9.4.10}$$

which involve two hypothetical source densities σ, ζ on ∂B, where θ has the same significance as in (4.5.13). Substituting from (9.4.7)–(9.4.10) into (9.4.6) and then into (9.4.3), we obtain two coupled boundary integral equations for σ, ζ to which a unique solution exists (Bhattacharyya, 1975). These source densities generate ϕ, $\bar{\phi}$, $\partial\psi/\partial x$, $\partial\psi/\partial y$, $\partial\phi/\partial x$, $\partial\phi/\partial y$—and therefore also $\partial\chi/\partial x$, $\partial\chi/\partial y$—in $B + \partial B$, whence $2\mu u$, $2\mu v$, $\partial^2\chi/\partial x^2$, $\partial^2\chi/\partial y^2$, $\partial^2\chi/\partial x\partial y$ may be generated in B. Also $\partial\chi/\partial x$, $\partial\chi/\partial y$ on ∂B immediately provide the traction components $\partial(\partial\chi/\partial x)/\partial q$, $\partial(\partial\chi/\partial y)/\partial q$ at any point of ∂B. There is no difficulty in adapting the formulation to mixed boundary-value problems of the type

$$\left.\begin{array}{l} 2\mu u,\ 2\mu v \text{ prescribed over } \partial B_d \\[6pt] \partial\chi/\partial x,\ \partial\chi/\partial y \text{ prescribed over } \partial B_t \end{array}\right\}, \quad \partial B = \partial B_d + \partial B_t. \tag{9.4.11}$$

We now construct biharmonic functions exhibiting given multi-valuedness in $\partial\chi/\partial x$, $\partial\chi/\partial y$, $\chi - x(\partial\chi/\partial x) - y(\partial\chi/\partial y)$ independently, no dislocations being present. Thus

$$\chi = \chi_1 = x\theta + My\log r = (1 - M)\,x\theta + M\,(x\theta + y\log r) \tag{9.4.12}$$

implies the jumps

$$\left[\frac{\partial\chi}{\partial x}\right] = 2\pi, \qquad \left[\frac{\partial\chi}{\partial y}\right] = 0, \qquad \left[\chi - x\frac{\partial\chi}{\partial x} - y\frac{\partial\chi}{\partial y}\right] = 0 \tag{9.4.13}$$

on circuiting the origin, with no dislocations present if

$$2(1 - v)(1 - M) - 1 = 0, \qquad \text{i.e. } M = (1 - 2v)/2(1 - v). \qquad (9.4.14)$$

Similarly

$$\chi = \chi_2 = y\theta - Nx \log r = (1 - N) y\theta - N (x \log r - y\theta) \quad (9.4.15)$$

implies the jumps

$$\left[\frac{\partial \chi}{\partial x}\right] = 0, \qquad \left[\frac{\partial \chi}{\partial y}\right] = 2\mu, \qquad \left[\chi - x\frac{\partial \chi}{\partial x} - y\frac{\partial \chi}{\partial y}\right] = 0, \qquad (9.4.16)$$

with no dislocations present if

$$2(1 - v)(1 - N) - 1 = 0, \qquad \text{i.e. } N = (1 - 2v)/2(1 - v). \quad (9.4.17)$$

Finally,

$$\chi = \chi_3 = \theta \qquad (9.4.18)$$

implies the jumps

$$\left[\frac{\partial \chi}{\partial x}\right] = 0, \qquad \left[\frac{\partial \chi}{\partial y}\right] = 0, \qquad \left[\chi - x\frac{\partial \chi}{\partial x} - y\frac{\partial \chi}{\partial y}\right] = 2\pi \qquad (9.4.19)$$

with no dislocations present. Reference to (9.3.2) shows that χ_1 provides a resultant force of amount -2π acting in the y direction, and that χ_2 provides a resultant force of amount 2π acting in the x-direction, upon any contour ∂B enclosing the origin. Also, reference to (9.3.5) shows that χ_3 provides a resultant couple of amount 2π acting upon ∂B. Formulae (9.4.3) readily give the displacement components $2\mu u_1$, $2\mu v_1$ and $2\mu u_2$, $2\mu v_2$ associated with χ_1 and χ_2, which agree with those already obtained directly for two-dimensional point forces of magnitude 2π (Section 8.2). Details of the proof are given by Bhattacharyya (1975). Finally we note that

$$2\mu u_3 = -\frac{\partial \theta}{\partial x} = \frac{\sin \theta}{r}, \qquad 2\mu v_3 = -\frac{\partial \theta}{\partial y} = -\frac{\cos \theta}{r} \qquad (9.4.20)$$

express the displacement components associated with χ_3.

If ∂B is the internal boundary of an infinite exterior domain, or one boundary of a ring-shaped domain, we replace the left-hand sides of equations (9.4.3) by

$$2\mu(u - \alpha u_1 - \beta u_2), \qquad 2\mu(v - \alpha v_1 - \beta v_2) \qquad (9.4.21)$$

to allow for the possibility of non-equilibrated tractions acting on ∂B. The two unknown coefficients α, β are compensated by the two side conditions

$$\int_{\partial B} \sigma(\mathbf{q})\, dq = 0, \qquad \int_{\partial B} \zeta(\mathbf{q})\, dq = 0 \qquad (9.4.22)$$

which ensure that

$$\left.\begin{array}{l} \phi, \dfrac{\partial \psi}{\partial x} = O(r^{-1}) \quad \text{as} \quad r \to \infty \\[4mm] \bar{\phi}, \dfrac{\partial \psi}{\partial y} = O(r^{-1}) \neq O(0) \quad \text{as} \quad r \to \infty \end{array}\right\}. \qquad (9.4.23)$$

These conditions eliminate the possibility of unwanted dislocations. Since, from (4.1.3),

$$\frac{\partial \psi}{\partial x} = \frac{\cos \theta}{r} \int_{\partial B} x\zeta(\mathbf{q})\, dq + \frac{\sin \theta}{r} \int_{\partial B} y\zeta(\mathbf{q})\, dq + O(r^{-2}) \quad \text{as} \quad r \to \infty,$$

we see that $\int_{\partial B} y\zeta(\mathbf{q})\, dq$ determines the weighting of χ_3 in χ. There is no difficulty in adapting the formulation to mixed boundary-value problems.

Part II

Applications

IBM Library order form

IBM THE LIBRARY, MAIL POINT 149
IBM UNITED KINGDOM
LABORATORIES LIMITED
HURSLEY PARK, WINCHESTER
HAMPSHIRE SO21 2JN

Order Number (00)	Date (OE)
6302	23.8.78

Author (20)

JASWON MA + SYMM GT

Title (00) COMPUTATIONAL MATHEMATICS AND
APPLICATIONS: INTEGRAL EQUATION
METHODS IN POTENTIAL THEORY AND

Bibliographic (30) ELASTOSTATICS

ACADEMIC PRESS

Requested by (OD)	Notes (OH)
B. CHASE	130901

0 12 381050 7

10

General Numerical Formulation

10.1 Introduction

In Part II we are concerned with the numerical solution of harmonic and biharmonic boundary-value problems as formulated in terms of scalar potential theory in Part I. For numerical applications of vector potential theory to problems of elasticity we refer the reader to the works of Rizzo (1967), Rizzo and Shippy (1968), Butterfield (1971), Barone and Robinson (1972), Cruse (1969, 1972a, b, 1973, 1974, 1975). Cruse and Rizzo (1975), Watson (1972) and Lachat and Watson (1976). For the use of boundary integral equations in conjunction with finite element methods see Cruse et al. (1976), Banerjee (1976), Wang and Blandford (1976) and Zienkiewicz et al. (1977). Whilst the choice of formulation for a particular problem is often governed by physical considerations, the solution is in general obtained by the following two-stage process. *First* the relevant integral equation, or system of two or more coupled integral equations, is solved numerically; *then* its solution is used to generate, by numerical quadrature, the required potential or biharmonic function or whatever other related function may be desired.

In this chapter we describe the basic numerical techniques employed. In subsequent chapters we give detailed analyses of the computational procedures and their application to a variety of problems, first in two and then in three dimensions.

10.2 Basic approximations

In both stages of the solution we are concerned with integrals of the form

$$\phi(\mathbf{p}) = \int_{\partial B} k(\mathbf{p}, \mathbf{q}) u(\mathbf{q}) \, dq, \qquad (10.2.1)$$

where ∂B is a closed contour (surface) in two (three) dimensions and, as in Part I, $\mathbf{p}$ and $\mathbf{q}$ are vector variables specifying points in the plane (in space) and on the contour (surface) ∂B respectively whilst dq denotes the arc (surface) differential at $\mathbf{q}$. The function $k(\mathbf{p}, \mathbf{q})$ is a known kernel whilst the function $u(\mathbf{q})$, which is often unknown (at least initially), may be either a simple-source density, a double-source density or, in Green's formula, a boundary value of the relevant harmonic function or of its normal derivative.

For the purposes of numerical computation, we divide ∂B into n smooth intervals (surface elements, in three dimensions) $\Delta_1, \Delta_2, \ldots, \Delta_n$, so that $\partial B = \Delta_1 + \Delta_2 \ldots + \Delta_n$, and approximate the function $u(\mathbf{q})$, $\mathbf{q} \in \partial B$, by a step-function

$$\tilde{u}(\mathbf{q}) = u_j, \qquad \mathbf{q} \in \Delta_j; \qquad j = 1, 2, \ldots, n, \qquad (10.2.2)$$

where the u_j are some constants. Correspondingly, we approximate the integral (10.2.1) by

$$\tilde{\phi}(\mathbf{p}) = \int_{\partial B} k(\mathbf{p}, \mathbf{q}) \tilde{u}(\mathbf{q}) \, dq \qquad (10.2.3)$$

which we write in the form

$$\tilde{\phi}(\mathbf{p}) = \sum_{j=1}^{n} u_j \int_j k(\mathbf{p}, \mathbf{q}) \, dq, \qquad (10.2.4)$$

where $\int_j$ denotes integration over the jth interval (surface element) Δ_j of ∂B. This approximation is adequate for a large variety of problems, as we shall see. However, for certain problems, e.g. for elastostatic problems formulated via vector potential theory, higher order approximations are preferable (Besuner and Snow, 1975; Lachat and Watson, 1975).

Formula (10.2.4) provides a simple numerical quadrature formula which is applicable to any integral of the form (10.2.1) with an integrable kernel. In particular, this formula is applicable to integrals involving the weakly singular kernels (2.2.1). The coefficients $\int_j k(\mathbf{p}, \mathbf{q}) \, dq$ can sometimes be evaluated analytically but at other times they must be approximated numerically.

The determination of these coefficients for the various kernels which arise in two and three dimensions will be discussed in Chapters 11 and 16 respectively. For the remainder of the present chapter we shall use two-dimensional terminology alone, with the understanding that for three dimensions we need only substitute "surface element" for "interval", "area" for "length" and so on.

10.3 Solution of the integral equations

Suppose now that the function u satisfies an integral equation of the form

$$\int_{\partial B} k(\mathbf{p}, \mathbf{q}) \, u(\mathbf{q}) \, dq + \alpha u(\mathbf{p}) = f(\mathbf{p}), \qquad \mathbf{p} \in \partial B, \tag{10.3.1}$$

where k is a known integrable kernel, α is a known constant and f is a given function. Then if we approximate u by a step-function $\tilde{u}$, as described above, we obtain, by virtue of (10.2.4),

$$\sum_{j=1}^{n} u_j \int_j k(\mathbf{p}, \mathbf{q}) \, dq + \alpha \tilde{u}(\mathbf{p}) = f(\mathbf{p}), \qquad \mathbf{p} \in \partial B. \tag{10.3.2}$$

To solve this equation we use the method of "collocation", or "point-matching", applying the equation at one particular point $\mathbf{q}_i$ (which we call a "nodal" point) in each interval Δ_i of ∂B. We thus obtain

$$\sum_{j=1}^{n} u_j \int_j k(\mathbf{q}_i, \mathbf{q}) \, dq + \alpha u_i = f_i, \qquad \mathbf{q}_i \in \Delta_i; \qquad i = 1, 2, \ldots, n, \tag{10.3.3}$$

where

$$f_i = f(\mathbf{q}_i). \tag{10.3.4}$$

In this way we approximate the integral equation (10.3.1) by a system of simultaneous linear algebraic equations, (10.3.3), in the constants u_j.

This method of discretising an integral equation is a particular case of the method of moments (Harrington, 1968), in which an approximate solution is sought in the form of an expansion

$$\tilde{u}(\mathbf{q}) = \sum_{j=1}^{n} u_j \psi_j(\mathbf{q}), \qquad \mathbf{q} \in \partial B, \tag{10.3.5}$$

in terms of certain "basis" functions

$$\psi_j(\mathbf{q}); \qquad j = 1, 2, \ldots, n, \tag{10.3.6}$$

and the integral equation is then satisfied approximately by taking moments with certain "testing" functions. If the testing functions are the same as the basis functions, this is known as Galerkin's method. In the present case the basis functions, corresponding to our definition (10.2.2) of $\tilde{u}$, are the "pulse" functions defined by

$$\psi_j(\mathbf{q}) = \begin{cases} 1, & \mathbf{q} \in \Delta_j, \\ 0, & \mathbf{q} \in \partial B - \Delta_j. \end{cases} \tag{10.3.7}$$

The testing functions, corresponding to collocation at the nodal points $\mathbf{q}_i$, are the Dirac delta functions

$$\delta(\mathbf{p} - \mathbf{q}_i); \qquad i = 1, 2, \ldots, n. \tag{10.3.8}$$

Thus, multiplying equation (10.3.2) by each of these delta functions in turn and integrating with respect to $\mathbf{p}$, we obtain equations (10.3.3).

Similarly, we may discretise any of the integral equations derived in Part I. We turn now, therefore, to the problem of solving such systems of equations as result. Note in passing that (10.3.5) is a type of finite element approximation (McDonald et al., 1973; Wait, 1974). Alternative methods for discretising the integral equations of potential theory are discussed by Kupradze and Aleksidze (1964) and by de Wolf and de Mey (1976).

10.4 Solution of simultaneous linear algebraic equations

The problem of solving a system of simultaneous linear algebraic equations is well documented elsewhere (e.g. Householder, 1964; Wilkinson, 1965;

Varga, 1962). Moreover, this problem is of such fundamental importance in numerical analysis that library programs are often available for its solution (e.g. Wilkinson and Reinsch, 1971). Here, therefore, we shall merely outline those methods which are most appropriate in the present context.

We consider first a system of n equations in n unknowns, which we write compactly, in matrix notation, as

$$\mathbf{Ax} = \mathbf{b}, \tag{10.4.1}$$

where $\mathbf{A} = (a_{ij})$ is an $n \times n$ matrix of known coefficients, $\mathbf{x} = (x_j)$ is an $n \times 1$ column vector of unknowns to be determined and $\mathbf{b} - (b_i)$ is a known $n \times 1$ column vector. For example, equations (10.3.3) are of this form, with

$$a_{ij} = \int_j k(\mathbf{q}_i, \mathbf{q}) \, dq + \alpha \delta_{ij}; \qquad i, j = 1, 2, \ldots, n, \tag{10.4.2}$$

$$x_j = u_j; \qquad j = 1, 2, \ldots, n, \tag{10.4.3}$$

and

$$b_i = f_i; \qquad i = 1, 2, \ldots, n. \tag{10.4.4}$$

We suppose further, at first, that the matrix $\mathbf{A}$ is non-singular, in which case equation (10.4.1) has a unique solution

$$\mathbf{x} = \mathbf{A}^{-1}\mathbf{b}, \tag{10.4.5}$$

where $\mathbf{A}^{-1}$ denotes the inverse of the matrix $\mathbf{A}$. This solution may be obtained, without computing the inverse matrix explicitly, by either a "direct" method or an "iterative" method.

For most of our problems we use a direct method such as the method of Gaussian elimination with row interchanges (partial pivoting), as described, for example, by Wilkinson (1965). This method is essentially a triangular decomposition method. If the equations (10.4.1), reordered according to the row interchanges, take the form

$$\tilde{\mathbf{A}}\mathbf{x} = \tilde{\mathbf{b}}, \tag{10.4.6}$$

then the matrix $\tilde{\mathbf{A}}$ may be factorised uniquely in the form

$$\tilde{\mathbf{A}} = \mathbf{LU}, \tag{10.4.7}$$

where L is a lower triangular matrix with unit diagonal elements and U is upper triangular. Introducing an auxiliary vector y, defined by

$$Ux = y, \qquad\qquad (10.4.8)$$

we thus obtain, from equations (10.4.6),

$$Ly = \tilde{b}. \qquad\qquad (10.4.9)$$

Hence, given $\tilde{b}$, we may immediately compute y by a forward substitution through equations (10.4.9) and the required solution x follows from equations (10.4.8) by a back-substitution. In the Gaussian elimination method, the factor L is never obtained explicitly, the equations (10.4.8) being determined by a process of elimination, as the name suggests. The computed solution is, of course, subject to rounding (truncation) errors and will not in general satisfy equations (10.4.1) exactly. However, provided that these equations are not too ill-conditioned†, the solution may be improved by a process of iterative refinement (National Physical Laboratory, 1961). A few iterations are usually sufficient to yield a solution which is correct to full machine accuracy. For our purposes, such accuracy is more than adequate in view of the approximate nature of our systems of equations. Since, moreover, the iterative procedure necessitates storage of the original matrix A, as well as its triangular form U, we therefore dispense with this refinement when n is large. Essentially this means that we apply the iterative refinement only to relatively simple problems in two dimensions, when n is unlikely to exceed 100.

For problems in three dimensions, when the number of equations to be solved may be so large that the high-speed working store of the computer cannot contain the whole of the matrix A, iterative methods (Varga, 1962) may be more easily applied than direct methods. We consider two such methods, the Jacobi and Gauss-Seidel iterative methods, for each of which we assume that the diagonal elements of A are all non-zero; if this is not so initially, it can be achieved by a simple rearrangement of the equations. We denote by D the diagonal matrix composed of these elements and by

$$E = L + U, \qquad\qquad (10.4.10)$$

† The equations (10.4.1) are said to be "ill-conditioned" if the matrix A is such that a small change in the vector b results in a large change in the solution x.

where $\mathbf{L}$ and $\mathbf{U}$ are strictly lower and upper triangular respectively, the remainder of the matrix $\mathbf{A}$; we have then

$$\mathbf{A} = \mathbf{D} + \mathbf{E} = \mathbf{D} + \mathbf{L} + \mathbf{U}. \tag{10.4.11}$$

With this notation, the Jacobi iterative method is defined by

$$\mathbf{D}\mathbf{x}^{(m+1)} = -\mathbf{E}\mathbf{x}^{(m)} + \mathbf{b}; \qquad m = 0, 1, 2, \ldots, \tag{10.4.12}$$

where $\mathbf{x}^{(0)}$ is an initial estimate of $\mathbf{x}$. In terms of components, this becomes

$$a_{ii} x_i^{(m+1)} = -\sum_{\substack{j=1 \\ j \neq i}}^{n} a_{ij}x_j^{(m)} + b_i; \qquad i = 1, 2, \ldots, n; \qquad m \geqslant 0, \tag{10.4.13}$$

whence, since $a_{ii} \neq 0$, the components of the $(m+1)$th iterate may be successively determined from those of the mth iterate. The Gauss-Seidel method is a natural extension of the Jacobi method, based upon the observation that, since the $x_i^{(m+1)}$ are evaluated successively by equations (10.4.13), each $x_j^{(m)}$ for $j < i$ may be replaced by its revised value $x_j^{(m+1)}$. Thus

$$a_{ii}x_i^{(m+1)} = -\sum_{j=1}^{i-1} a_{ij}x_j^{(m+1)} - \sum_{j=i+1}^{n} a_{ij}x_j^{(m)} + b_i; \qquad i = 1, 2, \ldots, n;$$

$$m \geqslant 0, \tag{10.4.14}$$

or, in matrix notation,

$$(\mathbf{D} + \mathbf{L})\mathbf{x}^{(m+1)} = -\mathbf{U}\mathbf{x}^{(m)} + \mathbf{b}; \qquad m = 0, 1, 2, \ldots. \tag{10.4.15}$$

Equations (10.4.12) and (10.4.15) may each be written in the form

$$\mathbf{x}^{(m+1)} = \mathbf{M}\mathbf{x}^{(m)} + \mathbf{c}; \qquad m = 0, 1, 2, \ldots, \tag{10.4.16}$$

where $\mathbf{M}$ is known as the iteration matrix associated with the original matrix $\mathbf{A}$. For the Jacobi method,

$$\mathbf{M} = -\mathbf{D}^{-1}\mathbf{E}, \tag{10.4.17}$$

whilst for Gauss-Seidel,

$$\mathbf{M} = -(\mathbf{D} + \mathbf{L})^{-1}\mathbf{U}. \tag{10.4.18}$$

The matrices $\mathbf{D}$ and $\mathbf{D} + \mathbf{L}$ are of course non-singular and easily inverted, being simply diagonal and lower triangular respectively, with no zero diagonal elements; the products of their inverses with the vector $\mathbf{b}$ give the two forms of the vector $\mathbf{c}$ in equations (10.4.16). As regards the convergence of this iterative scheme we note that, if the spectral radius† $\rho(\mathbf{M})$ of the matrix $\mathbf{M}$ is strictly less than unity, i.e. if

$$\rho(\mathbf{M}) < 1, \qquad\qquad\qquad\qquad (10.4.19)$$

then, for any initial estimate $\mathbf{x}^{(0)}$, the successive iterates defined by (10.4.16) converge to the unique solution of the equation

$$(\mathbf{I} - \mathbf{M})\mathbf{x} = \mathbf{c}, \qquad\qquad\qquad (10.4.20)$$

which is also the solution of the original equation (10.4.1). For the proof of this result and further details, concerning rates of convergence, etc., see again Varga (1962). Here we note only that if $\rho(\mathbf{M})$ is very near to unity, the convergence is likely to be very slow. In equation (10.4.20), $\mathbf{I}$ denotes the $n \times n$ identity matrix, with unit diagonal elements and zeros everywhere else. The uniqueness of the solution of this equation, i.e. the existence of the inverse of the matrix $\mathbf{I} - \mathbf{M}$, is a direct consequence of the assumed non-singularity of the matrix $\mathbf{A}$.

Suppose now that the $n \times n$ matrix $\mathbf{A}$ is singular, i.e. that $|\mathbf{A}| = 0$. In this case equations (10.4.1) have no solution unless they are consistent, for which a necessary and sufficient condition is that

$$r(\mathbf{A}) = r(\mathbf{A}|\mathbf{b}), \qquad\qquad\qquad (10.4.21)$$

where $r(\mathbf{A})$ denotes the rank of the matrix $\mathbf{A}$ and $r(\mathbf{A}|\mathbf{b})$ that of the "augmented" matrix formed by adding the extra column $\mathbf{b}$ to $\mathbf{A}$. Since $\mathbf{A}$ is singular,

$$r(\mathbf{A}) < n, \qquad\qquad\qquad\qquad (10.4.22)$$

---

† The "spectral radius" $\rho(\mathbf{M})$ of the matrix $\mathbf{M}$ is defined by

$$\rho(\mathbf{M}) = \max_{1 \leqslant i \leqslant n} |\lambda_i|,$$

where $\{\lambda_i\}$ are the eigenvalues of the matrix $\mathbf{M}$, i.e. those values of λ for which the equation

$$\mathbf{M}\mathbf{x} = \lambda\mathbf{x}$$

has non-trivial solutions (eigenvectors).

and when condition (10.4.21) does hold there is in fact an infinite number of solutions of equations (10.4.1). In particular, the homogeneous system

$$\mathbf{Ax} = 0 \qquad\qquad (10.4.23)$$

has an infinite number of non-trivial solutions, of which $n - r(\mathbf{A})$ are linearly independent. Arbitrary multiples of these solutions may be added to any solution of the inhomogeneous system. In cases which we shall encounter, $r(\mathbf{A}) = n - 1$ and equations (10.4.23) have only one independent non-trivial solution. It follows, when equations (10.4.1) are consistent, that at least two of these equations are linearly dependent. By replacing one of these dependent equations by an independent condition for uniqueness, we obtain a non-singular system of equations which we may solve as described previously.

To conclude this section we consider briefly the approximate solution of inconsistent and over- or under-determined systems of equations, which we write, as before, in the form (10.4.1) but with $\mathbf{A}$ now an $m \times n$ matrix and m not necessarily equal to n. We seek a solution of these equations in the least-squares sense, i.e. we seek a vector $\mathbf{x}$ which minimises the Euclidean length or "norm", defined by

$$\|\mathbf{d}\| = \left(\sum_{i=1}^{m} |d_i|^2 \right)^{1/2}, \qquad\qquad (10.4.24)$$

i.e.
/* $A x$ is the projection
of B onto the cols of A. */

of the residual vector

$$\mathbf{d} = \mathbf{Ax} - \mathbf{b}. \qquad\qquad (10.4.25)$$

/* by defn, $A^T d = 0$ */

This vector necessarily satisfies the "normal" equations

$$\boxed{\mathbf{A}^T\mathbf{Ax} = \mathbf{A}^T\mathbf{b},} \qquad\qquad (10.4.26)$$

where $\mathbf{A}^T$ denotes the transpose of the matrix $\mathbf{A}$. When $\mathbf{A}^T\mathbf{A}$ is non-singular, as when $m > n$ and $\mathbf{A}$ is of maximum rank n, equations (10.4.26) have a unique solution. This is *the* least-squares solution of equations (10.4.1) and may be computed as described, for example, by Householder (1964). More generally, $\mathbf{A}^T\mathbf{A}$ is singular and equations (10.4.26) have an infinite number of solutions of which we may seek that which has the minimum norm. This solution is unique and is given by

$$\mathbf{x} = \mathbf{Xb} \qquad\qquad (10.4.27)$$

where $\mathbf{X}$ is an $n \times m$ matrix known as the "pseudo-inverse" of the $m \times n$ matrix $\mathbf{A}$. This vector $\mathbf{x}$ is the "minimal least squares solution" of equation (10.4.1) and may be computed as described by Peters and Wilkinson (1970).

10.5 Generation of potentials and of their derivatives

Earlier in this chapter we showed how integral equations involving integrals of the form (10.2.1), such as arise in the formulation of boundary-value problems in Part I, are reduced to systems of simultaneous linear algebraic equations on applying quadrature formulae of the type (10.2.4). Suppose now that we have done this for a particular boundary-value problem and that we have solved the resulting system of equations by one of the methods of the previous section. Then, in general, we will have obtained a step-function approximation of the form (10.2.2) to each unknown boundary function appearing in the relevant integral equations; for example, a source density $u(\mathbf{q})$, $\mathbf{q} \in \partial B$, will now be approximated by constant values $u_1, u_2, \ldots, u_n$ in the intervals $\Delta_1, \Delta_2, \ldots, \Delta_n$, respectively, into which the boundary ∂B is divided.

How we proceed at this stage depends upon the precise nature of the boundary-value problem being considered. Occasionally we are required only to find the total source density

$$U = \int_{\partial B} u(\mathbf{q}) \, dq \tag{10.5.1}$$

on the boundary ∂B and this is readily approximated by

$$\tilde{U} = \sum_{j=1}^{n} u_j h_j, \tag{10.5.2}$$

where h_j denotes the length (or approximate length) of the interval Δ_j. More generally, however, we wish to determine values of the potential associated with this source density, i.e. typically

$$\phi(\mathbf{p}) = \int_{\partial B} k(\mathbf{p}, \mathbf{q}) u(\mathbf{q}) \, dq. \tag{10.5.3}$$

This is approximated by

$$\tilde{\phi}(\mathbf{p}) = \sum_{j=1}^{n} u_j \int_j k(\mathbf{p}, \mathbf{q}) \, dq \qquad (10.5.4)$$

in accordance with formula (10.2.4) and we say that the step-function $\{u_j | j = 1, 2, \ldots, n\}$ generates the potential $\tilde{\phi}$. Note that the function $k(\mathbf{p}, \mathbf{q})$ in expressions (10.5.3) and (10.5.4) need not be the same as the kernel involved in the integral equations; as in formula (10.2.1), k is used here to denote a general kernel, just as u has been used to represent any of a number of kinds of source density. Note also that the point $\mathbf{p}$ is no longer necessarily restricted to the boundary ∂B, but may lie anywhere in the domain B of the boundary-value problem; of course, some integral representations of the form (10.5.3), and their approximations of the form (10.5.4), are valid only on ∂B or only within B, but not both, and such cases must be clearly distinguished as they arise.

Whilst in many problems the computation of one or more potentials completes the solution, we are sometimes also required to determine derivatives of these potentials. For example, we may wish to evaluate $\partial \phi / \partial x$, for ϕ given by (10.5.3), at a point $\mathbf{p}$ with Cartesian coordinates (x, y). If $\mathbf{p} \in B$, we may generally differentiate beneath the integral sign in (10.5.3) to obtain

$$\frac{\partial \phi(\mathbf{p})}{\partial x} = \int_{\partial B} \frac{\partial k(\mathbf{p}, \mathbf{q})}{\partial x} u(\mathbf{q}) \, dq, \qquad \mathbf{p} \in B, \qquad (10.5.5)$$

which, by analogy with (10.5.4), we approximate by

$$\frac{\partial \tilde{\phi}(\mathbf{p})}{\partial x} = \sum_{j=1}^{n} u_j \int_J \frac{\partial k(\mathbf{p}, \mathbf{q})}{\partial x} \, dq, \qquad \mathbf{p} \in B. \qquad (10.5.6)$$

Thus, having obtained the u_j, we may compute $\partial \tilde{\phi} / \partial x$ as readily as $\tilde{\phi}$ itself. Similarly we may compute $\partial \tilde{\phi} / \partial y$ and corresponding approximations to higher derivatives of ϕ at points $\mathbf{p} \in B$. Note, however, that as the order of differentiation increases, the kernels become more strongly singular at $\mathbf{p} = \mathbf{q}$ on the boundary. Consequently, when $\mathbf{p} \in \partial B$, we cannot in general simply differentiate beneath the integral sign in formula (10.5.3) and the computation of derivatives becomes more difficult.

Suppose, for example, that ϕ is a simple-layer potential in two dimensions, so that (10.5.3) takes the form

$$\phi(\mathbf{p}) = \int_{\partial B} \log |\mathbf{p} - \mathbf{q}| \sigma(\mathbf{q}) \, dq. \tag{10.5.7}$$

Then we recall, from (4.1.6), that the normal derivative of ϕ at a point on the boundary is

$$\phi'(\mathbf{p}) = \int_{\partial B} \log' |\mathbf{p} - \mathbf{q}| \sigma(\mathbf{q}) \, dq + \pi\sigma(\mathbf{p}), \qquad \mathbf{p} \in \partial B, \tag{10.5.8}$$

where the last term arises from the singularity in the kernel. The tangential derivative of (10.5.7) remains continuous at the boundary (assumed smooth) and may be expressed in the form

$$\frac{\partial \phi(\mathbf{p})}{\partial s} = \int_{\partial B} \frac{\partial}{\partial s} \log |\mathbf{p} - \mathbf{q}| \sigma(\mathbf{q}) \, dq, \qquad \mathbf{p} \in \partial B, \tag{10.5.9}$$

provided that this integral is interpreted in the sense of the Cauchy principal value (Muskhelishvili, 1953; Mikhlin, 1957). It follows from (10.5.8) and (10.5.9) that

$$\frac{\partial \phi(\mathbf{p})}{\partial x} = \int_{\partial B} \frac{\partial}{\partial x} \log |\mathbf{p} - \mathbf{q}| \sigma(\mathbf{q}) \, dq + \pi\sigma(\mathbf{p})x'(\mathbf{p}), \qquad \mathbf{p} \in \partial B, \tag{10.5.10}$$

where $x'(\mathbf{p})$ is the direction cosine of the normal to ∂B at $\mathbf{p}$. We approximate (10.5.10) by

$$\frac{\partial \tilde{\phi}(\mathbf{p})}{\partial x} = \sum_{j=1}^{n} \sigma_j \int_{j} \frac{\partial}{\partial x} \log |\mathbf{p} - \mathbf{q}| \, dq + \pi\tilde{\sigma}(\mathbf{p})x'(\mathbf{p}), \qquad \mathbf{p} \in \partial B, \tag{10.5.11}$$

where $\tilde{\sigma}$ denotes the step-function approximation to σ. The computation of the coefficient of σ_j when $\mathbf{p} \in \Delta_j$ requires particular care in view of the Cauchy singularity in the kernel; this will be discussed further in the next chapter. The problems of computing higher derivatives of the simple-layer logarithmic potential have recently been studied by E. Jaswon (1973) with reference to biharmonic problems of thin plate theory; these will not be discussed further here.

10.6 Error estimation

In the numerical solution of harmonic and biharmonic boundary-value problems by integral equation methods, as described here, the main sources of error are as follows: ——

- (i) the approximation of each unknown boundary function by a step-function,
- (ii) the application of the resulting discretised integral equation at only a finite number of points,
- (iii) the approximation (where necessary) of the coefficients in the resulting system of simultaneous linear equations, and
- (iv) any further approximations made in generating the solution as described in the previous section.

Compared with these, the error arising from the solution of the linear equations is usually negligible and may therefore be ignored.

Clearly any error analysis which seeks to trace the accumulation of error as it arises from each of these sources is likely to be very complicated, if indeed it is possible at all. In general therefore, in the present work, no such analysis is attempted; instead a *posteriori* bounds on the error in the computed solution are estimated by more simple means.

Particularly useful in this respect is the maximum principle for harmonic functions (e.g. Weinberger, 1965) which states that a function which is harmonic in a bounded domain B has its maximum value, and indeed (since the negative of a harmonic is also harmonic) its maximum modulus, on the boundary ∂B. Thus, if $\tilde{\phi}$ is a computed harmonic function approximating a true potential ϕ in B, then, since $\tilde{\phi} - \phi$ is harmonic also, $|\tilde{\phi} - \phi|$ has its maximum value on the boundary ∂B. Hence, in the Dirichlet problem for Laplace's equation, when ϕ is given everywhere on ∂B, examination of $\tilde{\phi}$ on ∂B yields an upper bound on the modulus of its error throughout B. In practice, since it is not possible to evaluate $\tilde{\phi}$ at every point of ∂B, we obtain an estimate of the maximum error by comparing $\tilde{\phi}$ with ϕ at a finite number of points.

A similar, but less well-known, maximum principle is applicable to the problem with mixed boundary conditions

$$\left.\begin{array}{l} \alpha\phi + \phi_i' = \gamma_1 \text{ on } \partial B_1, \quad \alpha < 0, \\ \phi = \gamma_2 \text{ on } \partial B_2, \end{array}\right\} \tag{10.6.1}$$

where $\partial B_1 + \partial B_2 = \partial B$ as in (2.1.7). In this case we have

$$\phi \leqslant \max \{\max_{\partial B_1} (\gamma_1/\alpha), \max_{\partial B_2} \gamma_2\}, \qquad (10.6.2)$$

since, as before, ϕ attains its maximum value on ∂B. If this maximum occurs at a point $\mathbf{p}$, then (Protter and Weinberger, 1967)

$$\phi'_i(\mathbf{p}) < 0, \qquad (10.6.3)$$

whence, since $\alpha < 0$ in (10.6.1),

$$\phi < \gamma_1/\alpha \text{ on } \partial B_1. \qquad (10.6.4)$$

It follows that if $\tilde{\phi}$ approximates ϕ in B_i then

$$|\tilde{\phi} - \phi| < \max \{\max_{\partial B_1} |(\alpha\tilde{\phi} + \tilde{\phi}'_i - \gamma_1)/\alpha|, \max_{\partial B_2} |\tilde{\phi} - \gamma_2|\} \qquad (10.6.5)$$

for this problem.

When no maximum principle is available, we estimate the accuracy of our numerical solution by comparing results for various sub-divisions of the boundary and/or by solving test problems, with known solutions, using the same boundary sub-divisions. The latter procedure is recommended also as a safeguard against the possibility of minor programming errors passing unnoticed.

11

Two-Dimensional
Numerical Analysis

11.1 Boundary subdivision

The first step in the numerical solution of a boundary-value problem
formulated in terms of integral equations, is the subdivision of the boundary
∂B into suitably small sections. In two dimensions these sections are simply
intervals of smooth, or piecewise smooth, plane curves and we therefore
refer to the points of subdivision as "interval" points. This distinguishes
them from the "nodal" points, one within each interval, at which the integral
equations are applied. In Section 10.3 we introduced the notation $\mathbf{q}_j$ for the
nodal point in the interval Δ_j of ∂B; correspondingly, we let $\mathbf{q}_{j \pm \frac{1}{2}}$ denote the
interval points which bound this interval.

In choosing the interval points, it is important that we include any corners
of the boundary ∂B and also any points where a prescribed boundary condition changes. This ensures that each interval of the boundary is smooth,
so that the normal is well-defined at each nodal point, and it ensures that a
single boundary condition applies within each interval. Also, to make full use
of symmetry, it is desirable that intervals of ∂B should be symmetrically
distributed about any axes of symmetry, such axes intersecting ∂B at interval
rather than nodal points. We note that these restrictions on the placing of
the interval points may make it impossible to divide ∂B into intervals of
uniform length, and this is one reason why we use quadrature formula (10.2.4)
rather than any more sophisticated rule.

Having chosen the interval points on a given boundary ∂B, we must next
select the intermediate nodal points. If the interval Δ_j of ∂B is a straight

line, we invariably take its mid-point to be the nodal point in that interval, i.e.

$$\mathbf{q}_j = \tfrac{1}{2}(\mathbf{q}_{j-\frac{1}{2}} + \mathbf{q}_{j+\frac{1}{2}}).$$ (11.1.1)

In this case, the length of the interval is

$$h_j = |\mathbf{q}_{j+\frac{1}{2}} - \mathbf{q}_{j-\frac{1}{2}}|$$ (11.1.2)

and the normal, directed into the domain B, has direction cosines

$$x'_j = \cos(n, x) = (y_{j-\frac{1}{2}} - y_{j+\frac{1}{2}})/h_j,$$ (11.1.3)

$$y'_j = \cos(n, y) = (x_{j+\frac{1}{2}} - x_{j-\frac{1}{2}})/h_j,$$ (11.1.4)

where, typically, $x_{j+\frac{1}{2}}$ is the x-coordinate of the point $\mathbf{q}_{j+\frac{1}{2}}$. It is assumed here, and subsequently, that the boundary points are numbered in such a way that their subscripts increase when the boundary is described so as to keep the domain B on its left.

When the interval Δ_j is not a straight line, we may approximate it by a chord, in which case we again define the nodal point by formula (11.1.1) whilst expressions (11.1.2) to (11.1.4) become approximations to the interval length and the normal components. More generally, however, we select a nodal point $\mathbf{q}_j$ on the interval Δ_j of the boundary and approximate the length of this interval, if required, by

$$h_j = |\mathbf{q}_j - \mathbf{q}_{j-\frac{1}{2}}| + |\mathbf{q}_{j+\frac{1}{2}} - \mathbf{q}_j|.$$ (11.1.5)

This corresponds to approximating the boundary interval by the two chords which join its end-points to the nodal point within it as illustrated in Fig. 11.1.1. In this case, the components of the normal to ∂B at the point $\mathbf{q}_j$ are usually derived analytically from the equation of the boundary contour.

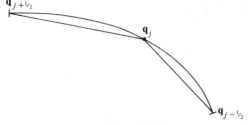

FIG. 11.1.1.

11.2 The kernel functions

Having divided the boundary ∂B into appropriate intervals and chosen the nodal points, our next step is to integrate the kernels of the relevant integral equations over the boundary intervals. In two dimensions, see Chapters 4 and 9, these kernels are made up of the following functions, either singly or in linear combination,

$$\log|\mathbf{p} - \mathbf{q}|, \tag{11.2.1}$$

$$\log'|\mathbf{p} - \mathbf{q}|, \tag{11.2.2}$$

$$\log|\mathbf{p} - \mathbf{q}|', \tag{11.2.3}$$

where the prime denotes differentiation along the normal to the boundary directed into the domain B at the point $\mathbf{p} \in \partial B$ in formula (11.2.2) and at the point $\mathbf{q} \in \partial B$ in formula (11.2.3). In each case $\mathbf{q}$ is the variable of integration; hence there is a significant difference between (11.2.2) and (11.2.3). As we shall see later, in Section 11.5, the kernel (11.2.3) may be integrated analytically over any smooth arc of ∂B for any position of the point $\mathbf{p}$, but this is not so of the kernel (11.2.2). If the point $\mathbf{p} \in \partial B$ has coordinates (x, y) and the normal to ∂B at this point has components (x', y'), then

$$\log'|\mathbf{p} - \mathbf{q}| = x'\frac{\partial}{\partial x}\log|\mathbf{p} - \mathbf{q}| + y'\frac{\partial}{\partial y}\log|\mathbf{p} \quad \mathbf{q}|. \tag{11.2.4}$$

Thus, for a "fixed" point $\mathbf{p}$, the kernel (11.2.2) is a linear combination, with "constant" coefficients, of the two kernels

$$\frac{\partial}{\partial x}\log|\mathbf{p} - \mathbf{q}|, \tag{11.2.5}$$

$$\frac{\partial}{\partial y}\log|\mathbf{p} - \mathbf{q}|. \tag{11.2.6}$$

The kernels (11.2.1), (11.2.5) and (11.2.6) appear not only in the integral equations themselves but also in formulae, such as (10.5.7) and (10.5.10), for the subsequent generation of logarithmic potentials and their first derivatives. In this context, they may refer to points $\mathbf{p}$ which are not on the boundary ∂B. At such points also we encounter, in the stress problems of Chapter 9, the second derivatives of the logarithmic kernel:

$$\frac{\partial^2}{\partial x^2} \log |\mathbf{p} - \mathbf{q}|, \qquad (11.2.7)$$

$$\frac{\partial^2}{\partial y^2} \log |\mathbf{p} - \mathbf{q}|, \qquad (11.2.8)$$

$$\frac{\partial^2}{\partial x \partial y} \log |\mathbf{p} - \mathbf{q}|. \qquad (11.2.9)$$

In considering the integration of these kernels, with respect to $\mathbf{q}$, we shall, for convenience, define

$$g = \log |\mathbf{p} - \mathbf{q}| \qquad (11.2.10)$$

and use the subscript notation for differentiation. Thus (11.2.5) becomes

$$g_x = \frac{\partial}{\partial x} \log |\mathbf{p} - \mathbf{q}|$$

or, carrying out the differentiation,

$$g_x = (x - x_q)/|\mathbf{p} - \mathbf{q}|^2, \qquad (11.2.11)$$

where $\mathbf{q} = (x_q, y_q)$. Similarly, (11.2.6) to (11.2.9) become

$$g_y = (y - y_q)/|\mathbf{p} - \mathbf{q}|^2, \qquad (11.2.12)$$

$$g_{xx} = 1/|\mathbf{p} - \mathbf{q}|^2 - 2(x - x_q)^2/|\mathbf{p} - \mathbf{q}|^4, \qquad (11.2.13)$$

$$g_{yy} = 1/|\mathbf{p} - \mathbf{q}|^2 - 2(y - y_q)^2/|\mathbf{p} - \mathbf{q}|^4, \qquad (11.2.14)$$

$$g_{xy} = -2(x - x_q)(y - y_q)/|\mathbf{p} - \mathbf{q}|^4. \qquad (11.2.15)$$

The above kernels are all singular at $\mathbf{p} = \mathbf{q}$. However, the singularity in g

is only a weak (integrable) singularity and the first derivatives g_x, g_y have at worst Cauchy singularities, which are integrable in the principal value sense. The second derivatives g_{xx}, g_{yy}, g_{xy} have stronger singularities which are not integrable in the normal sense.

11.3 Straight-line integration formulae

We consider first the case when the interval of integration is a straight line AB, Fig. 11.3.1, subtending an angle ψ at the point $\mathbf{p}$. Let the length of

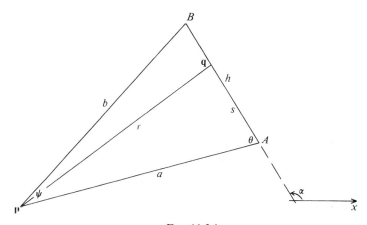

FIG. 11.3.1.

this line be h and let a and b be the distances of $\mathbf{p}$ from A and B respectively. Suppose that AB makes an angle α with the positive x-axis and let θ denote the angle $\mathbf{p}AB$. Provided that a, b and h are all non-zero, the angles ψ and θ are given by the formulae

$$\psi = \arccos\left(\frac{a^2 + b^2 - h^2}{2ab}\right), \qquad 0 \leqslant \psi \leqslant \pi, \qquad (11.3.1)$$

$$\theta = \arccos\left(\frac{a^2 + h^2 - b^2}{2ah}\right), \qquad 0 \leqslant \theta \leqslant \pi. \qquad (11.3.2)$$

For a general point $\mathbf{q}$ on AB, let the distance from $\mathbf{p}$, i.e. $|\mathbf{p} - \mathbf{q}|$, be denoted by

F

r and let the distance between A and $\mathbf{q}$ be s. Then

$$r^2 = s^2 + a^2 - 2as \cos \theta, \tag{11.3.3}$$

$$x_q = x_A + s \cos \alpha, \tag{11.3.4}$$

$$y_q = y_A + s \sin \alpha, \tag{11.3.5}$$

where x_A, y_A are the coordinates of the point A.

In this case the integral of kernel (11.2.10) is

$$\int_A^B g \, dq = \int_A^B \log |\mathbf{p} - \mathbf{q}| \, dq = I_1, \tag{11.3.6}$$

where

$$I_1 = \int_0^h \log r \, ds. \tag{11.3.7}$$

Using (11.3.4) we find that

$$\int_A^B g_x \, dq = (x - x_A)I_2 - \cos \alpha \, I_3, \tag{11.3.8}$$

where

$$I_2 = \int_0^h \frac{ds}{r^2}, \qquad I_3 = \int_0^h \frac{s \, ds}{r^2}. \tag{11.3.9}$$

Similarly

$$\int_A^B g_y \, dq = (y - y_A)I_2 - \sin \alpha \, I_3, \tag{11.3.10}$$

while, for kernels involving second-order derivatives, we obtain

$$\int_A^B g_{xx} \, dq = I_2 - 2(x - x_A)^2 I_4 + 4(x - x_A) \cos \alpha \, I_5 - 2 \cos^2\alpha \, I_6, \tag{11.3.11}$$

$$\int_A^B g_{yy}\, dq = I_2 - 2(y - y_A)^2 I_4 + 4(y - y_A)\sin \alpha\, I_5 - 2\sin^2\alpha\, I_6, \quad (11.3.12)$$

$$\int_A^B g_{xy}\, dq = - 2(x - x_A)(y - y_A)I_4$$

$$+ 2\{(x - x_A)\sin \alpha + (y - y_A)\cos \alpha\}I_5 - \sin 2\alpha\, I_6, \quad (11.3.13)$$

where

$$I_4 = \int_0^h \frac{ds}{r^4}, \qquad I_5 = \int_0^h \frac{s\, ds}{r^4}, \qquad I_6 = \int_0^h \frac{s^2\, ds}{r^4}. \quad (11.3.14)$$

Expressing r in terms of s by means of (11.3.3), we may in general evaluate the integrals $I_1, I_2, \ldots, I_6$ analytically. There are exceptional cases when $a = 0, b = 0$ or $\mathbf{p}, A$ and B are collinear, but if we ignore these (momentarily) we obtain

$$I_1 = a\cos \theta\,(\log a - \log b) + h(\log b - 1) + a\psi \sin \theta, \quad (11.3.15)$$

$$I_2 = \psi/(a\sin \theta), \quad (11.3.16)$$

$$I_3 = \log b - \log a + a\cos \theta\, I_2, \quad (11.3.17)$$

$$I_4 = \frac{1}{2a^2 \sin^2 \theta}\left\{\frac{h - a\cos \theta}{b^2} + \frac{\cos \theta}{a} + I_2\right\}, \quad (11.3.18)$$

$$I_5 = \tfrac{1}{2}\left(\frac{1}{a^2} - \frac{1}{b^2}\right) + a\cos \theta\, I_4, \quad (11.3.19)$$

$$I_6 = I_2 - a^2 I_4 + 2a\cos \theta\, I_5. \quad (11.3.20)$$

11.4 Straight-line integration formulae: special cases

When, in the notation of the previous section, $\mathbf{p}, A$ and B are collinear, $\theta = 0$ or π and therefore $\sin \theta = 0$ and $\cos \theta = 1$ or -1 respectively. In this case formula (11.3.15) may be simplified; for example, when $\theta = 0$

$$I_1 = a(\log a - 1) - b(\log b - 1). \quad (11.4.1)$$

Formula (11.4.1) reduces further, when $a = 0$ or $b = 0$, to

$$I_1 = h(\log h - 1), \tag{11.4.2}$$

from which it is evident that I_1 may be computed analytically for any position of the point $\mathbf{p}$.

When $\sin \theta = 0$, formula (11.3.16) for I_2 is either indeterminate, when $\psi = 0$ also, or it becomes infinite. In the former case, it is easily proved that

$$I_2 = \left| \frac{1}{b} - \frac{1}{a} \right|. \tag{11.4.3}$$

In the latter case, which arises when $\mathbf{p}$ lies in the interval AB, we note that, although I_2 and hence also I_3 become infinite, the integrals (11.3.8) and (11.3.10) may be given finite interpretations in the sense of the Cauchy principal value. For the configuration illustrated in Fig. 11.3.1, substitution of (11.3.16) and (11.3.17) into (11.3.8) and (11.3.10) yields

$$\int_A^B g_x \, dq = -\psi \sin \alpha - (\log b - \log a) \cos \alpha, \tag{11.4.4}$$

$$\int_A^B g_y \, dq = \psi \cos \alpha - (\log b - \log a) \sin \alpha. \tag{11.4.5}$$

These formulae are valid everywhere outside the interval AB; they also apply to the alternative configuration, where A and B are interchanged, provided the sign of the angle ψ is changed accordingly. For a point $\mathbf{p}$ on AB, we apply formulae (11.4.4) and (11.4.5) to intervals $(A, \mathbf{p} - \varepsilon)$ and $(\mathbf{p} + \varepsilon, B)$, whereby the Cauchy principal value analysis gives

$$\int_A^B g_x \, dq = (\log a - \log b) \cos \alpha, \qquad \mathbf{p} \in AB, \tag{11.4.6}$$

$$\int_A^B g_y \, dq = (\log a - \log b) \sin \alpha, \qquad \mathbf{p} \in AB. \tag{11.4.7}$$

In particular, when $\mathbf{p}$ is the mid-point of AB, so that $a = b$, these integrals reduce to zero. When $a = 0$ or $b = 0$, integrals (11.4.6) and (11.4.7) become infinite in magnitude, for a general value of α, in which case the Cauchy principal value analysis must be applied over the pair of adjacent boundary intervals meeting at the point $\mathbf{p}$. Exceptionally, expression (11.4.6) remains finite (actually zero) at A and B when $\cos \alpha = 0$, i.e. when $\alpha = \pi/2$ or $\alpha = 3\pi/2$, and similarly expression (11.4.7) remains finite (zero) when $\sin \alpha = 0$, i.e. when $\alpha = 0$ or $\alpha = \pi$. In each of these cases the integrand is the derivative normal to the line of AB, showing that in general it is the tangential component of the derivative which leads to the infinite behaviour. Note that when $\mathbf{p}$, A and B are collinear

$$\int_A^B \log' |\mathbf{p} - \mathbf{q}| \, dq = 0, \tag{11.4.8}$$

since the integrand is zero throughout the interval of integration. More generally this integral is obtained from results (11.4.4) and (11.4.5) following the decomposition (11.2.4).

When $\sin \theta = 0$, I_4, like I_2, becomes infinite unless $\psi = 0$, in which case

$$I_4 = \tfrac{1}{3} \left| \frac{1}{b^3} - \frac{1}{a^3} \right|. \tag{11.4.9}$$

When $\mathbf{p} \in AB$, I_4 and hence also I_5 and I_6 are all infinite in magnitude and in this case the singularities are so strong that integrals (11.3.11) to (11.3.13) cannot be evaluated even in the Cauchy principal value sense.

11.5 Further analytic integrations

We consider now the integration with respect to $\mathbf{q}$ of $\log |\mathbf{p} - \mathbf{q}|'$, i.e. kernel (11.2.3). This kernel can be integrated analytically not only along a straight line but along any smooth arc AB. Thus, recalling the Cauchy-Riemann equation (4.1.18), we have

$$\int_A^B \log |\mathbf{p} - \mathbf{q}|' \, dq = - \int_A^B \frac{\partial \theta (\mathbf{q} - \mathbf{p})}{\partial q} \, dq = - [\theta]_A^B, \tag{11.5.1}$$

where $[\theta]_A^B$ is the change in the angle $\theta(\mathbf{q} - \mathbf{p})$, which the vector $\mathbf{q} - \mathbf{p}$ makes with some fixed direction, as $\mathbf{q}$ moves from A to B. It is assumed here that θ increases in an anti-clockwise direction and that the arc AB is described so as to keep the domain into which the normal is directed on the left, as in Fig.

4.1.1. In a slightly more convenient notation, result (11.5.1) may be written

$$\int_A^B \log|\mathbf{p} - \mathbf{q}|' \, dq = -\theta_{AB, \mathbf{p}} \tag{11.5.2}$$

where $\theta_{AB, \mathbf{p}}$ is the angle subtended at the point $\mathbf{p}$ by the arc AB, with the appropriate sign attached, e.g. $\theta_{AB, \mathbf{p}} = \psi$ for the arrangement in Fig. 11.3.1.

If $\mathbf{p}$ lies on the arc AB, formula (11.5.2) requires slight modification, since the angle $\theta(\mathbf{q} - \mathbf{p})$ jumps by π as $\mathbf{q}$ passes through the point $\mathbf{p}$. This jump makes no contribution to the integral, which must in this case be considered in two parts. The integral from A to $\mathbf{p}$ contributes, by (11.5.2), the value $-\theta_{A\mathbf{p}, \mathbf{p}}$, where $\theta_{A\mathbf{p}, \mathbf{p}}$ is the signed angle between the straight line joining A to $\mathbf{p}$ and the tangent (directed as in Fig. 4.1.1) to the arc $A\mathbf{p}$ at $\mathbf{p}$. Similarly, the integral from $\mathbf{p}$ to B gives $-\theta_{\mathbf{p}B, \mathbf{p}}$, the negative of the signed angle between the tangent to the arc $\mathbf{p}B$ at $\mathbf{p}$ and the chord $\mathbf{p}B$. Adding these two results, we obtain

$$\int_A^B \log|\mathbf{p} - \mathbf{q}|' \, dq = -\theta_{A\mathbf{p}, \mathbf{p}} - \theta_{\mathbf{p}B, \mathbf{p}}, \qquad \mathbf{p} \in AB, \tag{11.5.3}$$

where the individual terms also cover the cases when $\mathbf{p}$ coincides with A or B.

A further analytic result to note when $\mathbf{p} \in AB$ is that if AB is a circular arc of radius a and length h, then

$$\int_A^B \log_i' |\mathbf{p} - \mathbf{q}| \, dq = \int_A^B \log|\mathbf{p} - \mathbf{q}|_i' \, dq = -h/2a, \tag{11.5.4}$$

since, for any two points $\mathbf{p}, \mathbf{q}$ on a circle of radius a,

$$\log_i' |\mathbf{p} - \mathbf{q}| = \log|\mathbf{p} - \mathbf{q}|_i' = -1/2a, \tag{11.5.5}$$

as is easily shown. Alternatively, the second of equalities (11.5.4) may be derived from (11.5.3), noting that the angle which AB subtends at the centre of the circle, namely h/a radians, is twice the angle which it subtends at the circumference. If the normal is directed outwards from the circle instead of inwards, results (11.5.5) becomes

$$\log_e' |\mathbf{p} - \mathbf{q}| = \log|\mathbf{p} - \mathbf{q}|_e' = 1/2a \tag{11.5.6}$$

and corresponding to (11.5.4) we have

$$\int_A^B \log_e' |\mathbf{p} - \mathbf{q}| \, dq = \int_A^B \log |\mathbf{p} - \mathbf{q}|_e' \, dq = h/2a. \tag{11.5.7}$$

This concludes our discussion of those cases, in two dimensions, for which we can integrate our kernels analytically.

11.6 Numerical quadrature

For intervals other than straight lines it is seldom possible to obtain the integrals we require analytically and some form of numerical quadrature must be used. Indeed, even for straight-line intervals it is not always necessary or desirable to carry out the integrations analytically. Analytic integration formulae are often more complicated and consequently take longer to compute than numerical quadrature formulae. Moreover, they do not necessarily lead to any greater accuracy in the results since this is limited by the basic step-function approximation—see, for instance, Example 13.3.1.

For any integrand $k(\mathbf{p}, \mathbf{q})$, the simplest quadrature formula for integration with respect to $\mathbf{q}$ over the interval Δ_j of length (or approximate length) h_j is

$$\int_j k(\mathbf{p}, \mathbf{q}) \, dq \doteqdot k(\mathbf{p}, \mathbf{q}_j) \, h_j, \tag{11.6.1}$$

where $\mathbf{q}_j$ is the nodal point within the interval. When $\mathbf{q}_j$ is the mid-point of Δ_j, formula (11.6.1) describes the mid-ordinate rule for integration. A more accurate result in this case, particularly if the integrand varies appreciably over the interval, is given by Simpson's rule

$$\int_j k(\mathbf{p}, \mathbf{q}) \, dq \doteqdot \{k(\mathbf{p}, \mathbf{q}_{j-\frac{1}{2}}) + 4k(\mathbf{p}, \mathbf{q}_j) + k(\mathbf{p}, \mathbf{q}_{j+\frac{1}{2}})\} h_j/6. \tag{11.6.2}$$

Though not strictly valid when $\mathbf{q}_j$ is not the mid-point of Δ_j, this formula may be applied in such cases with h_j the approximate length (11.1.5).

If, as is frequently the case in the present work, the integrand $k(\mathbf{p}, \mathbf{q})$ is singular, or undefined, when $\mathbf{p}$ and $\mathbf{q}$ coincide, then neither formula (11.6.1) nor (11.6.2) is applicable when $\mathbf{p}$ lies within the interval of integration. In this case we approximate the interval by one or two straight lines, or by a circular arc, and apply the analytic results of the previous sections. In particular, when $\mathbf{p} = \mathbf{q}_j$ we obtain, by reference to Fig. 11.1.1 and formula (11.4.2), the result

$$\int_j \log|\mathbf{q}_j - \mathbf{q}|\, dq \doteq |\mathbf{q}_j - \mathbf{q}_{j-\frac{1}{2}}|\,(\log|\mathbf{q}_j - \mathbf{q}_{j-\frac{1}{2}}| - 1)$$
$$+ |\mathbf{q}_{j+\frac{1}{2}} - \mathbf{q}_j|\,(\log|\mathbf{q}_{j+\frac{1}{2}} - \mathbf{q}_j| - 1). \qquad (11.6.3)$$

Similarly, when $\mathbf{p}$ coincides with one end of the interval of integration, i.e. when $\mathbf{p} = \mathbf{q}_{j\pm\frac{1}{2}}$, we obtain

$$\int_j \log|\mathbf{q}_{j\pm\frac{1}{2}} - \mathbf{q}|\, dq = |\mathbf{q}_{j+\frac{1}{2}} - \mathbf{q}_{j-\frac{1}{2}}|\,(\log|\mathbf{q}_{j+\frac{1}{2}} - \mathbf{q}_{j-\frac{1}{2}}| - 1) \qquad (11.6.4)$$

by approximating the interval Δ_j by a single chord. Also, if we approximate the interval by a circular arc, we obtain, from results (11.5.4) and (11.5.7),

$$\int_j \log'|\mathbf{q}_j - \mathbf{q}|\, dq \doteq \int_j \log|\mathbf{q}_j - \mathbf{q}|'\, dq = \pm \frac{h_j}{2a_j}, \qquad (11.6.5)$$

where a_j is the radius of the circle through the points $\mathbf{q}_{j-\frac{1}{2}}$, $\mathbf{q}_j$ and $\mathbf{q}_{j+\frac{1}{2}}$. Note that the equality on the right of (11.6.5) is consistent with result (11.5.3). The technique we have adopted here is sometimes referred to as integration over the singularity. Alternatively, one may extract the singularity as described by Christiansen (1971).

With reference to (11.6.5), we recall from (4.1.19) and (4.1.23) that, for a closed contour ∂B passing smoothly through a point $\mathbf{p}$,

$$\int_{\partial B} \log|\mathbf{p} - \mathbf{q}|'\, dq = \pm\pi, \qquad (11.6.6)$$

where the positive sign refers to the outward normal derivative and the negative sign to the inward normal derivative. Thus, if $\partial B = \sum_{j=1}^{n} \Delta_j$,

$$\int_j \log|\mathbf{q}_j - \mathbf{q}|'\, dq = \pm\pi - \sum_{\substack{i=1 \\ i \neq j}}^{n} \int_i \log|\mathbf{q}_j - \mathbf{q}|'\, dq, \qquad (11.6.7)$$

whence, applying quadrature rule (11.6.1) to each of the integrals on the right and bearing in mind (11.6.5), we obtain the alternative approximation

$$\int_j \log' |\mathbf{q}_j - \mathbf{q}| \, dq \doteq \pm\pi - \sum_{\substack{i=1 \\ i \neq j}}^{n} \log |\mathbf{q}_j - \mathbf{q}_i|' \, h_i. \qquad (11.6.8)$$

This approximation (with the negative sign) is particularly significant in the solution of the interior Neumann problem by means of the simple-layer potential as we shall see in Chapter 13.

11.7 Conjugate harmonic functions

To conclude this chapter, we consider the problem of computing the harmonic conjugate

$$\psi(\mathbf{p}) = \int_{\partial B} \theta(\mathbf{p} - \mathbf{q})\sigma(\mathbf{q}) \, dq \qquad (11.7.1)$$

of the simple-layer logarithmic potential

$$\phi(\mathbf{p}) = \int_{\partial B} \log |\mathbf{p} - \mathbf{q}| \sigma(\mathbf{q}) \, dq, \qquad (11.7.2)$$

where ∂B is an arbitrary closed contour. To this end, the generally multi-valued angle $\theta(\mathbf{p} - \mathbf{q})$, which the vector $\mathbf{p} - \mathbf{q}$ makes with some fixed direction, must be rigorously defined and we do this as follows:

(i) We let the fixed direction be that of the positive x-axis of a Cartesian coordinate system. Note that any change in this direction would introduce an additive constant into ψ, as appeared in formula (4.5.13).

(ii) We select a particular point $\mathbf{q}_0 \in \partial B$, to remain the same for all positions of the point $\mathbf{p}$, and define $\theta(\mathbf{p} - \mathbf{q}_0)$, positive in the anticlockwise sense, so that

$$-\pi < \theta(\mathbf{p} - \mathbf{q}_0) \leqslant \pi. \qquad (11.7.3)$$

If B is an exterior domain, we choose $\mathbf{q}_0$ so that all points $\mathbf{p}$ for which

$$\theta(\mathbf{p} - \mathbf{q}_0) = \pi \qquad (11.7.4)$$

lie within the domain B; i.e. we introduce a "cut" extending from $\mathbf{q}_0$ to infinity.

(iii) We let θ vary continuously, for a fixed point $\mathbf{p}$, as $\mathbf{q}$ describes the contour ∂B in the usual sense (i.e. keeping the domain B on the left) starting from the point $\mathbf{q}_0$.

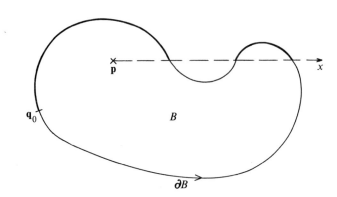

Fig. 11.7.1.

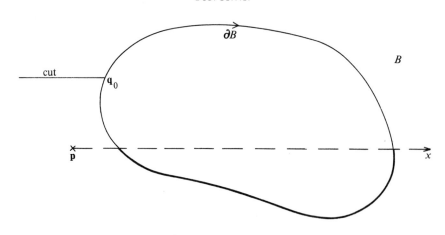

Fig. 11.7.2.

This definition of θ may be clarified by reference to Figs. 11.7.1 and 11.7.2 pertaining to interior and exterior domains respectively. Thus, for a given point $\mathbf{p}$, as indicated, and a general point $\mathbf{q} \in \partial B$, let $\vartheta(\mathbf{p} - \mathbf{q})$ denote the angle (between $\mathbf{p} - \mathbf{q}$ and the positive x-axis) in the range

$$-\pi < \vartheta(\mathbf{p} - \mathbf{q}) \leqslant \pi. \tag{11.7.5}$$

Then, from (11.7.3),

$$\theta(\mathbf{p} - \mathbf{q}_0) = \vartheta(\mathbf{p} - \mathbf{q}_0), \tag{11.7.6}$$

and it follows that

$$\theta(\mathbf{p} - \mathbf{q}) = \vartheta(\mathbf{p} -\!\!- \mathbf{q}) \tag{11.7.7}$$

whenever $\mathbf{q}$ lies on those parts of ∂B which are drawn thinly in the diagrams. On the thick parts of ∂B, on the other hand, we have

$$\theta(\mathbf{p} - \mathbf{q}) = \vartheta(\mathbf{p} - \mathbf{q}) + 2\pi \tag{11.7.8}$$

in Fig. 11.7.1 and

$$\theta(\mathbf{p} - \mathbf{q}) = \vartheta(\mathbf{p} - \mathbf{q}) - 2\pi \tag{11.7.9}$$

in Fig. 11.7.2.

When B is an interior domain, as in Fig. 11.7.1, $\theta(\mathbf{p} - \mathbf{q})$ increases by 2π as $\mathbf{q}$ describes a circuit of ∂B and θ therefore jumps by -2π as $\mathbf{q}$ passes through $\mathbf{q}_0$. With respect to $\mathbf{p}$, $\theta(\mathbf{p} - \mathbf{q})$ is continuous throughout B. When B is an exterior domain, as in Fig. 11.7.2, $\theta(\mathbf{p} - \mathbf{q})$ is continuous with respect to $\mathbf{q}$ for $\mathbf{p} \in B$ but θ jumps by 2π as $\mathbf{p}$ crosses the cut (11.7.4).

For a point $\mathbf{p} \in \partial B$, $\theta(\mathbf{p} - \mathbf{q})$ may be evaluated by a limiting process, by considering points within B as they approach the boundary. It will immediately be evident that θ jumps by π as $\mathbf{q}$ passes through $\mathbf{p}$, assuming that the boundary ∂B is smooth at $\mathbf{p}$. In the exceptional case when $\mathbf{p} = \mathbf{q}_0$, $\theta(\mathbf{p} - \mathbf{q}_0)$ is defined in terms of the tangent to ∂B at $\mathbf{q}_0$.

Returning now to the practical computation of ψ, we approximate (11.7.1), in the manner of (10.2.4), by

$$\tilde{\psi}(\mathbf{p}) = \sum_{j=1}^{n} \sigma_j \int_j \theta(\mathbf{p} - \mathbf{q}) \, dq. \tag{11.7.10}$$

In general θ is not integrable analytically and therefore we evaluate the coefficients of σ_j in (11.7.10) by numerical quadrature formulae of the form (11.6.1) or (11.6.2). It is convenient in this case to take $\mathbf{q}_0$, in the definition of θ, to be one of the interval points of the boundary subdivision.

12

Solution of Dirichlet Problems

12.1 Methods used

In this chapter we shall study the numerical solution of Dirichlet problems in two dimensions by methods based upon the simple-layer logarithmic potential as described in Section 4.3. This representation gives a more direct formulation of the Dirichlet problem, in terms of a Fredholm integral equation of the first kind, than does the classical double-layer potential representation, which gives rise to a Fredholm integral equation of the second kind.

It is well-known that a Fredholm equation of the first kind with a non-singular kernel can be very difficult to solve, being essentially ill-conditioned (Fox and Goodwin, 1953; Miller, 1974). Indeed, this fact may partly account for the past preference (noted in Section 2.6) for the double-layer potential formulation of Dirichlet problems, as applied numerically by, for example, Vlasov and Bakushchinskii (1963) or Kantorovich and Krylov (1964). However, in the present case, the singularity in the logarithmic kernel ensures diagonal dominance in the matrix of the corresponding algebraic system and the problem is in general well conditioned; the only exception is when the boundary ∂B is a Γ-contour (see Section 4.2) and this can always be avoided by scaling.

Suppose first that we wish to solve an interior Dirichlet problem, i.e. to obtain a harmonic function ϕ in a simply-connected domain B, bounded by a closed contour ∂B, given the values of ϕ everywhere on ∂B. Then, if we seek the solution of this problem in the form of a simple-layer potential

$$\phi(\mathbf{p}) = \int_{\partial B} \log |\mathbf{p} - \mathbf{q}| \sigma(\mathbf{q}) \, dq, \qquad \mathbf{p} \in B, \tag{12.1.1}$$

we find that the source density σ must satisfy the integral equation (4.3.1), i.e.

$$\int_{\partial B} \log |\mathbf{p} - \mathbf{q}| \sigma(\mathbf{q}) \, dq = \phi(\mathbf{p}), \qquad \mathbf{p} \in \partial B. \tag{12.1.2}$$

Discretising this equation, in the manner of Section 10.3, we obtain a system of n simultaneous linear equations:

$$\sum_{j=1}^{n} \sigma_j \int_j \log |\mathbf{q}_i - \mathbf{q}| \, dq = \phi(\mathbf{q}_i); \qquad i = 1, 2, \ldots, n, \tag{12.1.3}$$

for n unknowns $\sigma_1, \sigma_2, \ldots, \sigma_n$. Solving these equations directly, as described in Section 10.4, we approximate the solution ϕ by

$$\tilde{\phi}(\mathbf{p}) = \sum_{j=1}^{n} \sigma_j \int_j \log |\mathbf{p} - \mathbf{q}| \, dq, \qquad \mathbf{p} \in B + \partial B. \tag{12.1.4}$$

Similarly, if we recall the decomposition (4.3.3) and seek the solution of the given problem in the form

$$\phi(\mathbf{p}) - \int_{\partial B} \log |\mathbf{p} - \mathbf{q}| \sigma(\mathbf{q}) \, dq + k, \tag{12.1.5}$$

where k is a constant balanced by

$$\int_{\partial B} \sigma(\mathbf{q}) \, dq = 0, \tag{12.1.6}$$

we obtain the approximation

$$\tilde{\phi}(\mathbf{p}) = \sum_{j=1}^{n} \sigma_j \int_j \log |\mathbf{p} - \mathbf{q}| \, dq + \tilde{k}, \qquad \mathbf{p} \in B + \partial B. \tag{12.1.7}$$

In this case $\tilde{\sigma} \equiv \{\sigma_j | j = 1, 2, \ldots, n\}$ and $\tilde{k}$ may be obtained in either of two ways:

(i) Most simply, $\tilde{\sigma}$ and $\tilde{k}$ may be obtained by direct solution of the $n + 1$ simultaneous equations

$$\sum_{j=1}^{n} \sigma_j \int_j \log |\mathbf{q}_i - \mathbf{q}| \, dq + \tilde{k} = \phi(\mathbf{q}_i); \qquad i = 1, 2, \ldots, n, \qquad (12.1.8)$$

and

$$\sum_{j=1}^{n} \sigma_j h_j = 0, \qquad (12.1.9)$$

formed by discretising the boundary integral equation derived from (12.1.5) and the side condition (12.1.6).

(ii) Alternatively, $\tilde{k}$ may be obtained explicitly, from (4.3.4), in the form

$$\tilde{k} = \left(\sum_{j=1}^{n} \lambda_j \phi(\mathbf{q}_j) h_j \right) / \tilde{\kappa}, \qquad (12.1.10)$$

where

$$\tilde{\kappa} = \sum_{j=1}^{n} \lambda_j h_j \qquad (12.1.11)$$

approximates κ, (4.2.7), and $\tilde{\lambda} \equiv \{\lambda_j | j = 1, 2, \ldots, n\}$ satisfies

$$\sum_{j=1}^{n} \lambda_j \int_j \log |\mathbf{q}_i - \mathbf{q}| \, dq = 1; \qquad i = 1, 2, \ldots, n, \qquad (12.1.12)$$

i.e. the discrete form of equation (4.2.1). Then $\tilde{\sigma}$ may be obtained by solving equations (12.1.8) alone.

Whilst the second method of obtaining $\tilde{\sigma}$ and $\tilde{k}$ is a little more complicated than the first, it should be noted that the matrices of coefficients in equations (12.1.8) and (12.1.12) are identical. Thus solving the intermediate problem in λ involves very little extra effort; it may, moreover, provide useful supplementary information. For example, although the value of κ is seldom known analytically, a study of the convergence of $\tilde{\kappa}$ (to κ) as n is increased provides a check on the stability of the numerical procedure. Also, if ∂B has any symmetry, then λ and, for a suitable subdivision of the boundary (Section 11.1), $\tilde{\lambda}$ have the same symmetry and this provides a practical check on the matrix of coefficients common to systems (12.1.8) and (12.1.12). Note that σ will not be symmetric unless ϕ too is symmetric on ∂B.

For exterior Dirichlet problems, choice between the representations (12.1.1) and (12.1.5) is governed by the required behaviour at infinity (see Section 4.3). Also, for certain problems, such as that described in Section 12.7, further representations, which are extensions of the simple-layer potential, are appropriate. In each case, discretisation of the problem is essentially as described above. However, the methods may differ in the way in which the various coefficients of σ_j are computed. For convenience in the subsequent discussion, therefore, we categorise the methods as follows:

Method A: We use analytic integration over the two-chord approximation (Fig. 11.1.1) of each boundary interval, applying formula (11.3.6) in conjunction with (11.3.13), (11.4.1) and (11.4.2) as appropriate. The length (or approximate length) h_j of the interval Δ_j is given by formula (11.1.5).

Method B: We use numerical quadrature formula (11.6.2), in conjunction with analytic results (11.6.3) and (11.6.4) at the nodal and interval points respectively. When the boundary interval Δ_j is curved, h_j may be either

(i) the approximate length, given by formula (11.1.5), or
(ii) the true length of the curve.

The latter is seldom used and option (i) may be assumed unless otherwise stated.

12.2 A simple example

As our first illustration we consider a Dirichlet problem in a polygonal domain:

Example 12.2.1

The problem: To determine the harmonic function ϕ in the L-shaped domain shown in Fig. 12.2.1, given that $\phi = x$ on the boundary. This problem has of course the analytic solution $\phi = x$ throughout the bounded domain B.

The method: The modified simple-layer potential (12.1.5) is applied with the boundary ∂B divided into n equal intervals, each of length $h = 16/n$.

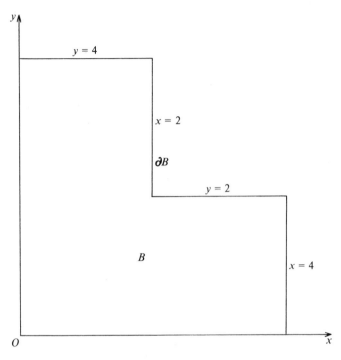

FIG. 12.2.1.

The solution is thus obtained in the form (12.1.7), where $\tilde{\sigma}$ and $\tilde{k}$ satisfy equations (12.1.8) and (12.1.9) with $\phi(\mathbf{q}_i) = x(\mathbf{q}_i)$ and $h_j = h$. The coefficients of σ_j in (12.1.7) and (12.1.9) are evaluated by Method B.

Results: Table 12.2.1 and Fig. 12.2.2 show computed results (Symm, 1963) for various values of n. In the Table, $\tilde{\kappa}$ comes from formula (12.1.11) and E denotes the maximum value of $|\tilde{\phi} - \phi|$ at the interval points, which, for each value of n considered, is the value at the corner $x = 4$, $y = 2$. The Figure shows the final subdivision ($n = 64$) of the boundary, together with the corresponding few computed values of $1000\,\tilde{\phi}$ at the interval points which differ by a unit or more from the values of $1000\,x$. The latter are included at the foot of the diagram.

Comments: It is interesting to observe that the re-entrant corner is not the source of the greatest error. At the corner where it does occur, the maximum error recorded when $n = 64$ is less than 3%. Away from the corners the errors rapidly diminish.

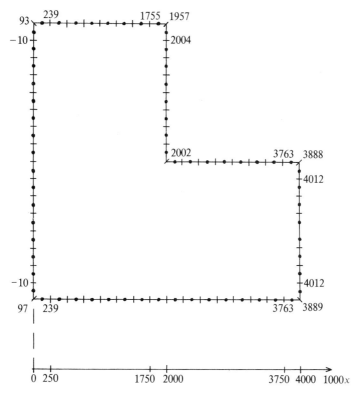

FIG. 12.2.2.

TABLE 12.2.1 L-shaped domain

n	$\tilde{k}$	$\tilde{\kappa}$	E
8	1·720	1·348	0·431
16	1·731	1·314	0·287
32	1·735	1·303	0·179
64	1·736	1·295	0·112

In view of the maximum principle (Section 10.6), E provides an estimate of the maximum error in $\tilde{\phi}$ throughout the domain B. We note that, since $\tilde{k}$ converges more rapidly (with n) than $\tilde{\phi}$ and $\tilde{\phi}(\mathbf{p}) \to \tilde{k}$ as $|\mathbf{p}| \to \infty$, E also provides an estimate of the maximum error in the bounded solution of the corresponding exterior problem.

An unbounded solution of the exterior problem may be determined by applying representation (12.1.2). In this case the solution $\tilde{\sigma}$ of equations (12.1.3) generates a slightly less accurate interior solution (12.1.4) than the one obtained here.

Since the boundary in this example is composed entirely of straight lines, all of the coefficients of σ_j could have been evaluated analytically using Method A. The reader may find this a useful exercise.

12.3 Corner singularities

It is well-known (Motz, 1946) that in the neighbourhood of a corner of angle α, of a boundary on which a harmonic function ϕ takes prescribed continuous values, the function ϕ may in general have the form

$$\phi = \phi_0 + \sum_{j=1}^{\infty} c_j r^{j\pi/\alpha} \sin\frac{j\pi\theta}{\alpha}, \qquad 0 \leqslant \theta \leqslant \alpha, \qquad (12.3.1)$$

where r, θ are polar coordinates relative to the corner, the c_j are constants and ϕ_0 has the prescribed values on $\theta = 0$ and $\theta = \alpha$. It follows, if $\alpha > \pi$, that the derivatives of ϕ may become infinite in magnitude as $r \to 0$, and it is not surprising, when such singularities arise, that it is difficult to compute ϕ accurately in this region.

In Example 12.2.1, it was noted that the errors near the corners pointing outwards were larger than those near the re-entrant corner. Though at first sight this may appear to contradict what has just been said, it is in fact a direct consequence of it. Since there is no singularity in the solution of the problem posed, all the c_j are zero in the expansion of this solution around the re-entrant corner. However, there are singularities in the solution of the corresponding exterior problem, at the corners which point into the exterior domain, whence, by virtue of formula (4.1.8), σ becomes infinite in magnitude at these corners. Consequently σ is not well approximated by a constant σ_j in an interval Δ_j adjacent to one of these corners and the solution $\tilde{\phi}$ suffers accordingly.

To illustrate this point more forcefully we consider the following exterior Dirichlet problem:

Example 12.3.1

The problem: To determine the harmonic function ϕ with logarithmic potential behaviour at infinity which takes the value $\phi = 1$ on the interval $-c \leqslant x \leqslant c$ of the x-axis, $y = 0$, when in particular $c = \sqrt{3}$. In general, i.e. for arbitrary c, this problem has the analytic solution (Symm, 1964)

$$\phi = 1 + \xi/\log(c/2), \tag{12.3.2}$$

where ξ is the confocal elliptic coordinate given by

$$\xi = \text{Re} \{\cosh^{-1}(z/c)\}, \qquad z = x + iy. \tag{12.3.3}$$

The method: In this example we apply the simple-layer potential (12.1.1) in the form

$$\phi(\mathbf{p}) - \int_{-c}^{c} \log |\mathbf{p} - \mathbf{x}| \lambda(x) \, dx, \tag{12.3.4}$$

where $\mathbf{x} = (x, 0)$. Thus, dividing the interval of integration into n equal subintervals, each of length $h = 2c/n$, we obtain an approximate solution

$$\tilde{\phi}(\mathbf{p}) = \sum_{j=1}^{n} \lambda_j \int_j \log |\mathbf{p} - \mathbf{x}| \, dx. \tag{12.3.5}$$

By choosing an even value for n and taking account of symmetry, we need to solve only $n/2$ simultaneous equations for the λ_j, the coefficients of which are evaluated by Method *B* with $h_j = h$.

Results: Table 12.3.1 shows a selection of computed results (Symm, 1963) for various values of n; $\tilde{\kappa}$ comes from formula (12.1.11) and E is the maximum error in ϕ at the interval points, viz. $|\tilde{\phi} - 1|$ at $x = \pm\sqrt{3}$, $y = 0$; $\lambda(x_j)$ is the analytic value of λ at the nodal point $\mathbf{x}_j = (x_j, 0)$, as derived from the formula

$$\lambda(x) = \frac{1}{\pi} \frac{\partial \phi}{\partial y} = [\pi \{\log(c/2)\}(c^2 - x^2)^{\frac{1}{2}}]^{-1}, \tag{12.3.6}$$

which follows from (12.3.2). From (12.3.6), or from (12.3.2) directly, we have also the analytic result

$$\kappa = \int_{-c}^{c} \lambda(x) \, dx = [\log(c/2)]^{-1}. \tag{12.3.7}$$

Parts (b) and (c) of the Table refer to the case when $n = 96$.

TABLE 12.3.1. Straight line ($c = \sqrt{3}$).

(a)	n		$\tilde{\kappa}$			E		
	16		$-6{\cdot}023$			$1{\cdot}167$		
	32		$-6{\cdot}454$			$0{\cdot}882$		
	64		$-6{\cdot}695$			$0{\cdot}646$		
	96		$-6{\cdot}777$			$0{\cdot}534$		
			$\kappa = -6{\cdot}952$					
(b) Interval point	$\frac{1}{2}$	$1\frac{1}{2}$	$2\frac{1}{2}$	$3\frac{1}{2}$	$4\frac{1}{2}$	$5\frac{1}{2}$		$96\frac{1}{2}$
$\tilde{\phi}$	$0{\cdot}466$	$1{\cdot}062$	$0{\cdot}999$	$1{\cdot}001$	$1{\cdot}000$	$1{\cdot}000$	...	$0{\cdot}466$
(c) Nodal point (j)	1	2	3	4	5	6		96
$-\lambda_j$	$11{\cdot}690$	$4{\cdot}996$	$4{\cdot}032$	$3{\cdot}395$	$3{\cdot}000$	$2{\cdot}721$	...	$11{\cdot}690$
$-\lambda(x_j)$	$8{\cdot}888$	$5{\cdot}147$	$4{\cdot}009$	$3{\cdot}410$	$3{\cdot}023$	$2{\cdot}749$	...	$8{\cdot}888$

Comments: Although $\tilde{\kappa} \to \kappa$ and $E \to 0$ as $n \to \infty$, the convergence in this example is slow on account of the singularities in $\lambda(x)$ at $x = \pm c$. Part (b) of the Table shows how badly these singularities affect the accuracy of $\tilde{\phi}$. Note, however, that the solution near each end of the straight-line boundary may be improved by taking account of the known analytic form of the local singularity. Thus, from formula (12.3.1) with $\phi_0 = 1$ and $\alpha = 2\pi$, we deduce that

$$\lambda(x) \sim k(x + c)^{-\frac{1}{2}} \text{ as } x \to - c, \tag{12.3.8}$$

where k is some constant. Since the inaccuracy of $\tilde{\phi}$ near $x = - c$ is due primarily to the inadequacy of the constant approximation to λ in the first interval of the boundary, assuming that we have numbered our intervals from left to right, we estimate the value of k in (12.3.8) by reference to the second interval. Hence, denoting this estimate by $\tilde{k}$, we have

$$\tilde{k} = \lambda_2(x_2 + c)^{\frac{1}{2}} = -1{\cdot}162 \text{ when } n = 96; \tag{12.3.9}$$

incidentally, from (12.3.6), the true value of k is

$$k = [\pi\{\log(c/2)\}(2c)^{\frac{1}{2}}]^{-1} = -1{\cdot}189. \tag{12.3.10}$$

Near $x = -c$, $y = 0$, therefore, we approximate ϕ by

$$\hat{\phi}(\mathbf{p}) = -1\cdot162 \int_1 (x + c)^{-\frac{1}{2}} \log |\mathbf{p} - \mathbf{x}| \, dx$$

$$+ \sum_{j=2}^{96} \lambda_j \int_j \log |\mathbf{p} - \mathbf{x}| \, dx. \qquad (12.3.11)$$

In particular, when $\mathbf{p} = (-c, 0)$ and we evaluate the first integral in (12.3.11) analytically and the remaining terms numerically as previously, we obtain $\hat{\phi} = 0\cdot993$. The latter shows a considerable improvement over the corresponding tabulated value $\tilde{\phi} = 0\cdot466$; the same improvement may be obtained at the point $(c, 0)$ in a similar manner.

Note that whilst we have described here an *a posteriori* method for improving the results in Table 12.3.1, an *a priori* method of incorporating the analytic behaviour of λ into the solution of this problem could have been derived along similar lines (cf. Chapter 14).

12.4 Conformal mapping

In Section 4.5 we formulated problems of conformal mapping in terms of the simple-layer logarithmic potential and its conjugate. We now illustrate the numerical application of this formulation with reference to a doubly-connected domain.

Example 12.4.1

The problem: To determine the function $w(z)$, $z = x + iy$, which maps the doubly-connected domain B, bounded externally by a circle ∂B_0 with equation

$$x^2 + y^2 = a^2 \qquad (12.4.1)$$

and internally by a square ∂B_1 given by

$$-b \leqslant x \leqslant b, \qquad -b \leqslant y \leqslant b, \qquad (12.4.2)$$

conformally onto a circular annulus in such a way that ∂B_0 and ∂B_1 pass to the outer and inner circumferences respectively. Since the ratio of the radii of these circumferences is unique (Kantorovich and Krylov, 1964), the outer radius may be taken to be unity and the inner radius denoted by $\rho < 1$, where ρ has to be determined.

The method: Since $z = 0$ lies inside the domain bounded by ∂B_1, the mapping function may be expressed in the form (4.5.16). Correspondingly, we obtain numerically the mapping function

$$\tilde{w}(z) = z\, e^{\tilde{\phi} + i\tilde{\psi}}, \tag{12.4.3}$$

where $\tilde{\phi}$ and $\tilde{\psi}$ approximate conjugate harmonic functions in B. Dividing ∂B_0 and ∂B_1 into n_0 and n_1 equal intervals respectively, we represent $\tilde{\phi}$ by (12.1.4) with $n = n_0 + n_1$, $\partial B = \partial B_0 + \partial B_1$. Equations (4.5.17) and (4.5.18) then yield $n + 1$ simultaneous linear equations for $\tilde{\sigma}$ and $\log \tilde{\rho}$, where $\tilde{\rho}$ approximates ρ. The coefficients of σ_j in these equations and in the expression (12.1.4) for $\tilde{\phi}$ are evaluated by Method B. In terms of $\tilde{\sigma}$, $\tilde{\psi}$ is given by

$$\tilde{\psi}(\mathbf{p}) = \sum_{j=1}^{n} \sigma_j \int_j \theta(\mathbf{p} - \mathbf{q})\, dq \tag{12.4.4}$$

which is evaluated as described in Section 11.7. In particular, $\tilde{\phi}$ and $\tilde{\psi}$, and hence $\tilde{w}$, are evaluated at the interval points of ∂B. The coefficients of σ_j in (12.4.4) are evaluated by numerical quadrature formula (11.6.2) except at $\mathbf{p} = \mathbf{q}_{j + \frac{1}{2}}$, where (11.6.1) is used.

In practice, taking n_0 and n_1 and hence n to be multiples of four, we pre-serve the symmetry of the problem and reduce the number of equations solved to $n/4 + 1$. Since the mapping is unique only to within an arbitrary rotation, a constant may be added to $\tilde{\psi}$ to make the real axes correspond.

Results: Table 12.4.1 illustrates results obtained for three values of a with b fixed. In part (a), E denotes the maximum absolute deviation at the interval points of ∂B of $|\tilde{w}|$ from either 1, on ∂B_0, or $\tilde{\rho}$, on ∂B_1. In each case this turns out to be $||\tilde{w}| - \tilde{\rho}|$ at the corner of the square.

TABLE 12.4.1. Square in circle: doubly-connected mapping ($b = 5$)

(a) Values of E

$a \backslash n$	64	128	256	512
7·7	0·0527	0·0323	0·0197	0·0122
9·4	0·0363	0·0211	0·0128	0·0080
10·0	0·0336	0·0195	0·0119	0·0074

(b) Values of $\tilde{\rho}$

$a \backslash n$	64	128	256	512
7·7	0·7737	0·7782	0·7797	0·7803
9·4	0·6261	0·6283	0·6291	0·6294
10·0	0·5880	0·5900	0·5907	0·5910

Comments: As n increases, E decreases, though the convergence is slow on account of the corners in ∂B_1, particularly for small a when these corners lie close to ∂B_0. However, $|\,|\tilde{w}|\, - \tilde{\rho}|$ decreases rapidly away from the corners and $\tilde{\rho}$, which varies little with n, appears to be accurate, for the final subdivision, to within 0·001.

To corroborate this claim we note that the exterior of the square ∂B_1 can be mapped analytically (Savin, 1961) onto the exterior of a circle γ_1, the circle ∂B_0 being simultaneously mapped onto a near-circular contour γ_0. Then applying the method of the present example to the resulting doubly-connected domain, whose boundaries γ_0 and γ_1 contain no corners, we find that, for $n = 256$, E is of order 10^{-4} and $\tilde{\rho}$ takes the values 0·7806, 0·6296 and 0·5911 in the three cases considered. Though the problem has no known analytic solution, Laura (1965) and Gaier (1964) provide estimates 0·7843, 0·6285 and 0·5911 for ρ in these cases. Also, in the final case, $a = 10·0$, Ugodčikov (1955) gives a value 0·5922 and Opfer (1967) gives an upper bound of 0·6005 for ρ.

In view of the accuracy of $\tilde{\rho}$, it follows from the maximum principle (Section 10.6) that E provides an estimate of the error in the computed mapping function. Ideally, of course, we should compare $|\tilde{w}|$ with ρ on ∂B_1 but this is not possible since ρ is unknown. That $\tilde{\rho}$ is invariably more accurate than $\tilde{w}$ is evident from further examples of doubly-connected mappings (Symm, 1969a). For examples of the mapping of the interior or the exterior of a simply-connected domain see Symm (1966, 1967). For further refinements of the method see Hayes *et al.* (1972, 1975) and Gaier (1976).

12.5 Error bounds

In the previous examples, and in earlier work (Symm, 1963), we have taken E, the maximum value of $|\tilde{\phi} - \phi|$ at the interval points of ∂B, to be an estimate of the error in the approximate solution $\tilde{\phi}$ throughout the domain B. Note, however, that if a strict upper bound on the error is required, $\tilde{\phi}$ should be computed at many more points of ∂B, as the following simple example indicates.

Example 12.5.1

The problem: To determine the harmonic function ϕ in the domain B bounded by the circle ∂B:

$$x^2 + y^2 = 4, \tag{12.5.1}$$

given that $\phi = x$ on ∂B. This problem has the obvious analytic solution $\phi = x$ throughout B.

The method: The simple-layer potential (12.1.1) is applied with ∂B divided into n equal intervals. The solution is thus obtained in the form (12.1.4), where $\tilde{\sigma}$ satisfies equations (12.1.3) with $\phi(\mathbf{q}_i) = x(\mathbf{q}_i)$. In this example the coefficients of σ_j in (12.1.3) and (12.1.4) are evaluated by both Method A and Method B for comparison. Taking n to be a multiple of 4 and preserving the symmetry of the problem, we need to solve only $n/4$ equations in each case.

Results: Table 12.5.1 shows a selection of computed results. In part (a), the subscript A or B attached to $\tilde{\phi}$ indicates the method of integration used in its evaluation. In part (b), E_1 is the maximum absolute error in $\tilde{\phi}$ at the interval points of ∂B whilst E_2 denotes the maximum error in $\tilde{\phi}$ evaluated every quarter-interval of ∂B, i.e. not only at the interval and nodal points but also midway between these.

TABLE 12.5.1. Circular domain.

(a) $n = 32$				
x	1.0000	1.5000	1.9900	1.9999
$\tilde{\phi}_A(x, 0)$	1.0007	1.5011	1.9902	1.9994
$\tilde{\phi}_B(x, 0)$	0.9995	1.4993	2.0169	2.2084

(b) Method A			
n	$\tilde{\phi}(1, 0)$	E_1	E_2
8	1.0059	0.0336	—
16	1.0024	0.0041	0.0174
32	1.0007	0.0005	0.0043

Comments: It is evident from part (a) of the Table that if ϕ is required at points in B very close to the boundary, e.g. near the point (2,0), then $\tilde{\phi}_A$ provides a much better approximation than $\tilde{\phi}_B$. This is not surprising in view of the singularity in the logarithmic kernel on the boundary, where we integrate it analytically in any case.

Part (b) of the Table shows how important it is that the maximum absolute error in $\tilde{\phi}$ at the interval points, in this case denoted by E_1, should not be regarded as a strict upper bound on the over-all error in $\tilde{\phi}$ but only as an estimate. Indeed, we see that when $n = 32$, the error in $\tilde{\phi}$ at the point (1, 0), and also, from part (a), at the point (1·5, 0), actually exceeds E_1. Investigation of $\tilde{\phi}$ at many more points of ∂B suggests that E_2, as evaluated above, is much nearer to an 'upper bound' on the error in $\tilde{\phi}$.

12.6 Poisson's equation

Boundary-value problems for Poisson's equation may be reduced to similar problems for Laplace's equation by subtracting out a particular solution, independent of the boundary conditions.

Example 12.6.1

The problem: To solve Poisson's equation

$$\nabla^2 w = -1 \tag{12.6.1}$$

in the square domain B: $-1 < x, y < 1$, given that

$$w = 0 \tag{12.6.2}$$

on the boundary ∂B: $x, y = \pm 1$.

A particular solution of equation (12.6.1) is $-r^2/4$, $r^2 = x^2 + y^2$, since $\nabla^2(r^2) = 4$. In this case, therefore, we let

$$w = (\phi - r^2)/4 \tag{12.6.3}$$

and seek the function ϕ which satisfies

$$\nabla^2 \phi = 0 \text{ in } B \tag{12.6.4}$$

with the boundary condition

$$\phi = r^2 \text{ on } \partial B. \tag{12.6.5}$$

The method: Applying the simple-layer logarithmic potential representation (12.1.1), we obtain the approximate solution $\tilde{\phi}$, (12.1.4), to the problem defined by equations (12.6.4) and (12.6.5). Hence, from (12.6.3), we obtain

$$\tilde{w} = (\tilde{\phi} - r^2)/4 \tag{12.6.6}$$

as the approximate solution of Poisson's equation (12.6.1) with boundary condition (12.6.2).

In this example, in which the boundary is composed entirely of straight lines, each coefficient of σ_j, in expression (12.1.4) for $\tilde{\phi}$, is obtained analytically by Method A. The boundary is divided into n equal intervals and symmetry is taken into account to reduce the number of linear equations from n to $n/8$.

Results: Table 12.6.1 compares computed values of $\tilde{w}$, (12.6.6), for $n = 64$ and $n = 128$, denoted by $\tilde{w}_{64}$ and $\tilde{w}_{128}$ respectively, with corresponding approximations w_T, obtained by Tottenham (1970), at four points inside the domain B.

TABLE 12.6.1. Square: Poisson's equation.

x	y	$\tilde{w}_{64}$	$\tilde{w}_{128}$	w_T
0·00	0·00	0·2947	0·2947	0·2947
0·00	0·50	0·2293	0·2293	0·2294
0·50	0·50	0·1811	0·1811	0·1812
0·75	0·75	0·0726	0·0728	0·0729

Comments: The function w in this example may be interpreted physically, when multiplied by an appropriate constant, as either the torsion function for a prismatic bar of square cross-section (Timoshenko and Goodier, 1951) or as the deflection in certain circumstances of a thin square plate with simply-supported edges (Timoshenko and Woinowsky-Kreiger, 1959). Problems of these types will be considered in detail in later examples.

12.7 A practical problem

To conclude this chapter we consider the problem of computing the potential in a so-called multi-wire proportional counter. Typically, such a counter consists of a hollow cylinder of uniform cross-section at zero potential, through which pass a number of thin wires of uniform circular cross-section at unit potential. The wires run parallel to the cylinder and both are assumed to be infinitely long so that the potential satisfies Laplace's equation in two dimensions in any plane cross-section which cuts the cylinder orthogonally.

Let ∂B_0 denote the outer boundary of the counter and suppose that there are m internal wires. In general, the radii $R_1, R_2, \ldots, R_m$ of these wires are so small compared with their spacing and with the dimensions of ∂B_0 that their circular boundaries $\partial B_1, \partial B_2, \ldots, \partial B_m$ may be treated essentially as single points. Consequently, we seek the potential ϕ in the form

$$\phi(\mathbf{p}) = \int_{\partial B_0} \log |\mathbf{p} - \mathbf{q}| \sigma(\mathbf{q}) \, dq + \sum_{k=1}^{m} A_k \log |\mathbf{p} - \mathbf{w}_k|, \qquad \mathbf{p} \in B, \qquad (12.7.1)$$

where the A_k are constants (representing the charge on each wire) and $\mathbf{w}_k$ is the position vector of the centre of the kth wire.

Dividing ∂B_0 into n intervals, we discretise the integral in expression (12.7.1) in the usual manner. Then the boundary condition $\phi = 0$ on ∂B_0 yields the equations

$$\sum_{j=1}^{n} \sigma_j \int_j \log |\mathbf{q}_i - \mathbf{q}| \, dq + \sum_{k=1}^{m} A_k \log |\mathbf{q}_i - \mathbf{w}_k| = 0; \qquad i = 1, 2, \ldots, n.$$

$$(12.7.2)$$

Similarly, neglecting the thickness of each wire (on which $\phi = 1$) relative to its distance from any other wire or from ∂B_0, we obtain

$$\sum_{j=1}^{n} \sigma_j \int_j \log |\mathbf{w}_s - \mathbf{q}| \, dq + \sum_{\substack{k=1 \\ k \neq s}}^{m} A_k \log r_{ks} + A_s \log R_s = 1;$$

$$s = 1, 2, \ldots, m \qquad (12.7.3)$$

where r_{ks} denotes the distance $|\mathbf{w}_k - \mathbf{w}_s|$ between the centres of the kth and sth wires. Solving the $n + m$ equations (12.7.2) and (12.7.3) for σ and $A_1, A_2, \ldots,$ A_m, we thus obtain the approximate potential

$$\tilde{\phi}(\mathbf{p}) = \sum_{j=1}^{n} \sigma_j \int_j \log |\mathbf{p} - \mathbf{q}| \, dq + \sum_{k=1}^{m} A_k \log |\mathbf{p} - \mathbf{w}_k| \qquad (12.7.4)$$

within the counter.

Example 12.7.1
The problem: To determine the potential ϕ in a semicircular counter of radius 2·0 cm containing two wires of radius 0·002 cm with centres 1·2 cm apart, each 1·0 cm from the centre of the semicircle. The outer boundary of the counter is at zero potential and each wire is at unit potential.

The method: Applying the representation (12.7.1) to ϕ, with $m = 2$, we obtain the approximate solution $\tilde{\phi}$ in the form (12.7.4). The coefficients of σ_j are evaluated by Method A. Symmetry is taken into account to reduce the number of equations solved from $n + 2$ to $n/2 + 1$.

Results: Table 12.7.1 shows the computed values of the constant A_1 $(= A_2)$ for $n = 20$ and $n = 40$. The semicircle and straight line which bound the counter are each divided into equal intervals, the numbers of which are in the ratio $3:2$. Fig. 12.7.1 shows equipotential contours drawn mechanically

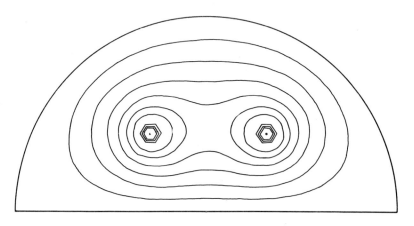

FIG. 12.7.1.

(Heap, 1972) from values of $\tilde{\phi}$ computed at points of a triangular mesh in the case when $n = 20$. The contours represent potential values 0.05, 0.10, 0.15, 0.20, 0.25, 0.30, 0.40, 0.50 and 0.60 moving from the boundary at zero potential towards the wires at unit potential.

TABLE 12.7.1.

n	$A_1 (= A_2)$
20	-0.152623
40	-0.152626

Comments: The hexagonal form of the contours near each wire is due to the use of straight lines to join points of equal potential on the sides of the triangles, after linear interpolation along these sides. For further examples of this nature see Symm (1974).

13

Solution of Neumann Problems

13.1 General formulation

In this chapter we consider the numerical solution of Neumann problems in two dimensions by means of the simple-layer logarithmic potential. Thus, given the normal derivative ϕ' on ∂B of a harmonic function ϕ in B, we seek ϕ in the form

$$\phi(\mathbf{p}) = \int_{\partial B} \log|\mathbf{p} - \mathbf{q}|\sigma(\mathbf{q})\,\mathrm{d}q, \qquad \mathbf{p} \in B + \partial B. \qquad (13.1.1)$$

We assume here that ϕ' denotes the normal derivative directed into the domain B. Then, by virtue of (4.1.6) or (4.1.7) according as B is an interior or exterior domain respectively, the source density σ must satisfy the integral equation

$$\int_{\partial B} \log'|\mathbf{p} - \mathbf{q}|\sigma(\mathbf{q})\,\mathrm{d}q + \pi\sigma(\mathbf{p}) = \phi'(\mathbf{p}), \qquad \mathbf{p} \in \partial B. \qquad (13.1.2)$$

As in the previous chapter, we approximate expression (13.1.1) by

$$\tilde{\phi}(\mathbf{p}) = \sum_{j=1}^{n} \sigma_j \int_j \log|\mathbf{p} - \mathbf{q}|\,\mathrm{d}q, \qquad \mathbf{p} \in B + \partial B, \qquad (13.1.3)$$

in which we evaluate the coefficients of σ_j by Method A or Method B (Section

175

12.1). Correspondingly, we approximate equation (13.1.2) by

$$\sum_{j=1}^{n} \sigma_j \int_j \log' |\mathbf{q}_i - \mathbf{q}| \, dq + \pi\sigma_i = \phi'(\mathbf{q}_i); \qquad i = 1, 2, \ldots, n, \qquad (13.1.4)$$

and we subsequently categorise methods for solving the Neumann problem according to the way in which we evaluate the coefficients of σ_j in these equations.

13.2 Methods for interior problems

We recall from Section 2.4 that the interior Neumann problem has a solution if, and only if,

$$\int_{\partial B} \phi'(\mathbf{p}) \, dp = 0 \qquad (13.2.1)$$

and that this solution is unique only to within an arbitrary additive constant. Correspondingly, equation (13.1.2) has a solution only if condition (13.2.1) holds, and this solution is unique only to within an arbitrary additive multiple of λ— a solution of the homogeneous equation

$$\int_{\partial B} \log' |\mathbf{p} - \mathbf{q}| \lambda(\mathbf{q}) \, dq + \pi\lambda(\mathbf{p}) = 0, \qquad \mathbf{p} \in \partial B. \qquad (13.2.2)$$

Provided that ∂B is not a Γ-contour (Section 4.2), λ may be identified with the unique solution of the equation

$$\int_{\partial B} \log |\mathbf{p} - \mathbf{q}| \lambda(\mathbf{q}) \, dq = 1, \qquad \mathbf{p} \in \partial B. \qquad (13.2.3)$$

Then λ generates a unit potential throughout B and adding $k\lambda$ to σ in (13.1.2), where k is any constant, is equivalent to adding k to ϕ. If ∂B is a Γ-contour, equation (13.2.3) has no solution, but any solution of equation (13.2.2) satisfies

$$\int_{\partial B} \log |\mathbf{p} - \mathbf{q}| \lambda(\mathbf{q}) \, dq = 0, \qquad \mathbf{p} \in B + \partial B. \qquad (13.2.4)$$

In this case, adding $k\lambda$ to σ does not change ϕ but we can still add an arbitrary constant to ϕ itself.

Suppose now that condition (13.2.1) is satisfied by a given $\phi'(\mathbf{p})$, $\mathbf{p} \in \partial B$. Then, to obtain a unique solution of equation (13.1.2) we must impose some extra 'normalising' condition upon σ, e.g. either

$$\int_{\partial B} \sigma(\mathbf{q}) \, dq = c \qquad (13.2.5)$$

or

$$\sigma(\mathbf{p}) = c \text{ for some } \mathbf{p} \in \partial B, \qquad (13.2.6)$$

where c is a prescribed constant. (We could prescribe $c = 0$ except in the case of a homogeneous system.) In practice, we discretise condition (13.2.5) in the form

$$\sum_{j=1}^{n} \sigma_j h_j = c, \qquad (13.2.7)$$

whilst, identifying $\mathbf{p}$ with the nodal point $\mathbf{q}_n$, equation (13.2.6) becomes

$$\sigma_n = c. \qquad (13.2.8)$$

In (13.2.7), h_j represents the length of the interval Δ_j which, as in the previous chapter, is usually given by the two-chord approximation (11.1.5). Coupling either (13.2.7) or (13.2.8) with (13.1.4), we have $n + 1$ equations relating the n unknowns $\sigma_1, \sigma_2, \ldots, \sigma_n$. We consider here two methods for obtaining an approximation $\tilde{\sigma} = \{\sigma_j | j = 1, 2, \ldots, n\}$ to σ from these equations.

Method C: Applying numerical quadrature formulae (11.6.1) and (11.6.8), we approximate equations (13.1.4) by

$$\tilde{\mathbf{A}} \boldsymbol{\sigma} = \boldsymbol{\phi}', \qquad (13.2.9)$$

where the elements of the matrix $\tilde{\mathbf{A}}$ are given by

$$\left. \begin{aligned} \tilde{a}_{ij} &= h_j \log' |\mathbf{q}_i - \mathbf{q}_j|, \qquad i \neq j, \\ \tilde{a}_{jj} &= -\sum_{\substack{i=1 \\ i \neq j}}^{n} h_i \log' |\mathbf{q}_i - \mathbf{q}_j|, \end{aligned} \right\} \qquad (13.2.10)$$

and the ith element of the vector ϕ' is $\phi'(\mathbf{q}_i)$. In this case, the matrix $\tilde{\mathbf{A}}$ is singular since, from (13.2.10),

$$\sum_{i=1}^{n} h_i a_{ij} = 0; \qquad j = 1, 2, \ldots, n, \qquad (13.2.11)$$

i.e. the rows of $\tilde{\mathbf{A}}$ are linearly dependent. Indeed, it follows from (13.2.11) that each row of $\tilde{\mathbf{A}}$ is a linear combination of the other $n - 1$ rows of the matrix. However, if we choose our subdivision of ∂R in such a way that

$$\sum_{i=1}^{n} \phi'(\mathbf{q}_i) h_i = 0, \qquad (13.2.12)$$

corresponding to equation (13.2.1), it follows from (13.2.11) and (13.2.12) that equations (13.2.9) are consistent. Then any $n - 1$ of these equations, coupled with the normalising condition (13.2.7) or (13.2.8), may be solved directly for $\sigma_1, \sigma_2, \ldots, \sigma_n$.

Method D: If each interval Δ_j of ∂B is a straight line (or if we approximate Δ_j by the chord joining the interval points $\mathbf{q}_{j \pm \frac{1}{2}}$), we may evaluate the coefficients of σ_j in equations (13.1.4) analytically by means of formulae (11.2.4), (11.4.4), (11.4.5) and (11.4.8). We thus obtain

$$\hat{\mathbf{A}}\boldsymbol{\sigma} = \boldsymbol{\phi}', \qquad (13.2.13)$$

where, in particular, the diagonal elements of $\hat{\mathbf{A}}$ are given by

$$\hat{a}_{jj} = \pi; \qquad j = 1, 2, \ldots, n. \qquad (13.2.14)$$

In this case, the matrix $\hat{\mathbf{A}}$ is not generally singular but the equations (13.2.13) are ill-conditioned when n is large. We therefore couple these equations with the normalising condition (13.2.7) and solve the resulting system of $n + 1$ equations in n unknowns by the method of least squares (Section 10.4).

13.3 The homogeneous problem

The interior Neumann problem with the homogeneous boundary condition

$$\phi'(\mathbf{p}) = 0, \qquad \mathbf{p} \in \partial B, \qquad (13.3.1)$$

has the obvious analytic solution

$$\phi(\mathbf{p}) = k, \qquad \mathbf{p} \in B + \partial B, \qquad (13.3.2)$$

where k is an arbitrary constant. Nevertheless, as a preliminary to solving more general problems, it is of interest to construct a solution of this problem by the numerical techniques of the previous section.

Clearly the boundary condition (13.3.1) satisfies the Gauss condition (13.2.1), and the corresponding integral equation

$$\int_{\partial B} \log' |\mathbf{p} - \mathbf{q}| \sigma(\mathbf{q}) \, dq + \pi\sigma(\mathbf{p}) = 0, \qquad \mathbf{p} \in \partial B, \qquad (13.3.3)$$

has a solution which is unique but for an arbitrary multiplying factor. A particular non-trivial solution σ of this equation may be sought by imposing condition (13.2.5) or (13.2.6) with $c \neq 0$.

Subsequently, the potential generated by this source density may either

(i) be scaled by a constant factor

or (ii) have a constant added to it

to give a potential with a prescribed value at a specified point of $B + \partial B$. Since option (i) is particular to the homogeneous problem whilst option (ii) is applicable in general, we adopt option (ii) in the example which follows.

Example 13.3.1

The problem: To determine the harmonic function ϕ in the square domain $B: -a \leqslant x, y \leqslant a$, given that $\phi' = 0$ everywhere on the boundary ∂B and $\phi = 1$ at the origin. The analytic solution is $\phi = 1$ throughout B.

The method: Bearing in mind option (ii) above, we seek ϕ in the form

$$\phi(\mathbf{p}) = \int_{\partial B} \log |\mathbf{p} - \mathbf{q}| \sigma(\mathbf{q}) \, dq + d, \qquad \mathbf{p} \in B + \partial B, \qquad (13.3.4)$$

where d is a constant to be determined. Then σ satisfies equation (13.3.3), which we discretise and solve by (a) Method C and (b) Method D, dividing ∂B into n equal intervals in each case. Having thus obtained $\tilde{\sigma}$, we approximate ϕ by

$$\tilde{\phi}(\mathbf{p}) = \sum_{j=1}^{n} \sigma_j \int_j \log |\mathbf{p} - \mathbf{q}| \, dq + \tilde{d}, \qquad \mathbf{p} \in B + \partial B, \qquad (13.3.5)$$

G

in which we evaluate the integrals by Method A of Chapter 12 and obtain $\tilde{d}$, approximating d, by setting $\tilde{\phi}(\mathbf{p}) = 1$ at $x = 0$, $y = 0$.

Results: Table 13.3.1 shows the errors E in $\tilde{\phi}$ at the corners of the square of side 4 ($a = 2$) for various values of n. These are the maximum errors in the solution evaluated every quarter interval of the boundary. The subscripts C and D indicate the methods of solution.

TABLE 13.3.1 Square ($a = 2$)

n	E_C	E_D
8	0·127	0·127
16	0·087	0·100
32	0·057	0·074
64	0·036	0·054

Comments: The choice of the square domain for this example enables us to apply Method D without having to approximate the boundary. Unfortunately, at the corners of such a domain, the source density σ (proportional to λ) becomes infinite and is therefore not well represented by a step-function; consequently, the errors tabulated are rather large (over 3 %). However, away from the corners the errors rapidly diminish and when $n = 64$ and Method C is applied, for example, $\tilde{\phi}$ is accurate to within 0·1 % at one interval length from each corner. On account of the symmetry of this problem, there are only $n/8$ independent σ_j and the number of equations solved may be reduced accordingly. Method C is slightly more accurate than Method D.

13.4 The general interior problem

As in the preceding example, we find in general that Method C, when applicable, is more accurate than Method D. We therefore use Method C whenever the boundary can be subdivided in such a way that condition (13.2.12) holds.

Example 13.4.1

The problem: To determine the harmonic function ϕ in the domain B

bounded by the circle ∂B:

$$(x - a)^2 + (y - a)^2 = a^2, \tag{13.4.1}$$

given that

$$\phi'(\mathbf{p}) = (x - y)/(x^2 + y^2), \qquad \mathbf{p} = (x, y) \in \partial B, \tag{13.4.2}$$

and that

$$\phi = \pi/4 \quad \text{at} \quad x = a, \qquad y = a. \tag{13.4.3}$$

This problem has the analytic solution

$$\phi(\mathbf{p}) = \text{arc tan}(y/x), \qquad \mathbf{p} = (x, y) \in B + \partial B, \tag{13.4.4}$$

which is independent of the radius a of ∂B.

The method: Whether we seek ϕ in the form (13.1.1) or (13.3.4), the source density σ satisfies equation (13.1.2) with ϕ' given by (13.4.2). Since ϕ' is anti-symmetric, whilst ∂B is symmetric, about the line $y = x$, we divide ∂B into an even number n of equal intervals symmetrically placed about this line. Then condition (13.2.12) holds automatically, with

$$h_i = h = 2\pi a/n, \qquad i = 1, 2, \ldots, n, \tag{13.4.5}$$

and we may discretise equation (13.1.2) by Method C. Here we solve the resulting system of linear equations (13.2.9) in two ways:

(i) Applying representation (13.1.1) to ϕ and assuming, with no loss of generality, that $a \neq 1$, we obtain, from (13.4.3),

$$\int_{\partial B} \sigma(\mathbf{q}) \, dq = \pi/(4 \log a). \tag{13.4.6}$$

Discretising this, we obtain the normalising condition

$$\sum_{j=1}^{n} \sigma_j = n/(8a \log a), \tag{13.4.7}$$

of the form (13.2.7). Coupling this with any $n - 1$ of equations (13.2.9), we solve directly for $\tilde{\sigma}$.

(ii) Applying the normalising condition (13.2.8), we set $\sigma_n = 0$ in any $n - 1$ of equations (13.2.9) and solve directly for $\sigma_1, \sigma_2, \ldots, \sigma_{n-1}$.

In case (i), $\tilde{\sigma}$ generates an approximation to ϕ of the form (13.1.3) and we denote this solution by $\tilde{\phi}_{(i)}$. In case (ii), we obtain an approximation to ϕ of the form (13.3.5) with

$$\tilde{d} = \pi/4 - 2\pi a \log a \sum_{j=1}^{n-1} \sigma_j/n \qquad (13.4.8)$$

and we denote this solution by $\tilde{\phi}_{(ii)}$. In each case, we evaluate the approximate solution by Method B.

Results: Table 13.4.1 and Fig. 13.4.1 show computed results (Symm, 1964) for various values of n and $a = 2$. In the Table, $E_{(i)}$ and $E_{(ii)}$ denote the maximum values of $|\tilde{\phi}_{(i)} - \phi|$ and $|\tilde{\phi}_{(ii)} - \tilde{\phi}|$ respectively at the points shown

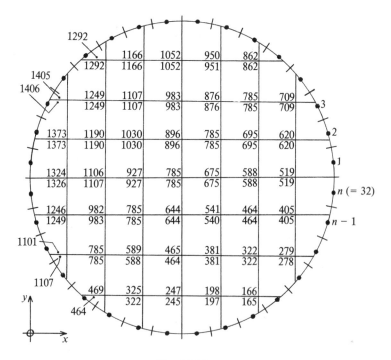

FIG. 13.4.1.

TABLE 13.4.1. Circle ($a = 2$)

n	$E_{(i)}$	$E_{(ii)}$
4	0·366	0·362
8	0·051	0·048
16	0·019	0·016
32	0·006	0·006

in the Figure. The Figure itself shows the final subdivision of the boundary and corresponding values of 1000 $\tilde{\phi}_{(n)}$, above each line, with values of 1000 ϕ, below the line, for comparison.

Comments: The alternative normalising conditions applied in this example yield results which differ relatively little. The solution illustrated in Fig. 13.4.1 compares favourably with the finite difference solutions of Fox (1944) and of Allen and Robins (1962).

13.5 Methods for exterior problems

The exterior Neumann problem has a unique solution with logarithmic potential behaviour at infinity. Correspondingly, if we seek this solution in the form (13.1.1), the integral equation (13.1.2) has a unique solution for arbitrary prescribed values of the outward normal derivative ϕ'. In particular, when $\phi' = 0$ on ∂B, $\sigma \equiv 0$ on ∂B and the homogeneous problem has only the trivial solution $\phi \equiv 0$ in $B + \partial B$.

In general, the solution σ of equation (13.1.2) satisfies

$$2\pi \int_{\partial B} \sigma(\mathbf{q}) \, dq = \int_{\partial B} \phi'(\mathbf{p}) \, dp, \qquad (13.5.1)$$

as we recall from (4.1.9). Hence, it is desirable that the solution $\tilde{\sigma}$ of the corresponding discrete equations (13.1.4) should satisfy

$$2\pi \sum_{j=1}^{n} \sigma_j h_j = \sum_{i=1}^{n} \phi'(\mathbf{q}_i) \, h_i. \qquad (13.5.2)$$

If the coefficients of σ_j in equations (13.1.4) are evaluated analytically, the solution will not usually satisfy this condition. We therefore adopt the following procedure:

Method E: Applying numerical quadrature formulae (11.6.1) and (11.6.8), we approximate equations (13.1.4) by

$$\bar{\mathbf{A}}\boldsymbol{\sigma} = \boldsymbol{\phi}', \tag{13.5.3}$$

where the elements of the matrix $\bar{\mathbf{A}}$ are given by

$$\left.\begin{aligned} \bar{a}_{ij} &= h_j \log'|\mathbf{q}_i - \mathbf{q}_j|, \qquad i \neq j, \\ \bar{a}_{jj} &= 2\pi - \sum_{\substack{i=1 \\ i \neq j}}^{n} h_i \log'|\mathbf{q}_i - \mathbf{q}_j|. \end{aligned}\right\} \tag{13.5.4}$$

In this case,

$$\sum_{i=1}^{n} h_i \bar{a}_{ij} = 2\pi h_j \tag{13.5.5}$$

and result (13.5.2) follows automatically. Equations (13.5.3) are solved directly for $\tilde{\sigma}$, which generates the approximate solution $\tilde{\phi}$ by (13.1.3).

As described in Section 10.5, the solution $\tilde{\sigma}$ also generates approximations $\tilde{\phi}_x, \tilde{\phi}_y$ to the first derivatives of ϕ with respect to Cartesian coordinates. This will be illustrated in the next section with reference to a problem from the theory of aerodynamics.

13.6 Flow past an obstacle

As in three dimensions (Section 2.4) the integral equation for the exterior Neumann problem, in this case equation (13.1.2), has an immediate application in the theory of potential fluid motion. Thus, suppose that ∂B is the boundary of a rigid obstacle which perturbs a steady flow defined by a velocity potential ψ. Then, if ϕ is the perturbation potential, so that the resultant velocity potential is $\phi + \psi$, the physical requirement that the component of fluid velocity normal to the rigid boundary must be zero implies that

$$\phi'(\mathbf{p}) = -\psi'(\mathbf{p}), \qquad \mathbf{p} \in \partial B. \tag{13.6.1}$$

Consequently, ϕ satisfies an exterior Neumann problem and, if we seek ϕ in the form (13.1.1), the source density σ satisfies equation (13.1.2). Note that from (13.5.1) and (13.6.1)

$$\int_{\partial B} \sigma(\mathbf{q}) \, dq = 0, \tag{13.6.2}$$

analogous with (2.4.13), whence from (4.1.3)

$$\phi(\mathbf{p}) = O(|\mathbf{p}|^{-2}) \quad \text{as} \quad |\mathbf{p}| \to \infty. \tag{13.6.3}$$

Hence the simple-layer potential representation for ϕ ensures that the perturbation of the fluid flow decays to zero sufficiently rapidly at infinity.

Example 13.6.1

The problem: To determine the pressure coefficient

$$C_p = 1 - (|\mathbf{V}|/|\mathbf{U}|)^2 \tag{13.6.4}$$

on an aerofoil in a stream of incompressible, inviscid fluid moving with constant velocity U at infinity. It may be assumed here that

(i) there is no circulation around the aerofoil, and therefore no lift acting upon it, and that
(ii) the velocity field is irrotational.

The method: The physical assumptions (i) and (ii) are such that the above theory applies with ∂B denoting the boundary of the aerofoil. With no loss of generality we let

$$\psi = -x, \tag{13.6.5}$$

so that

$$\mathbf{U} = -\nabla \psi = (1, 0), \quad |\mathbf{U}| = 1. \tag{13.6.6}$$

Then, since the velocity V of the fluid around the aerofoil is given by

$-\nabla(\phi + \psi)$, we have

$$C_p = 1 - \{(\phi_x - 1)^2 + \phi_y^2\} = 2\phi_x - \phi_x^2 - \phi_y^2. \tag{13.6.7}$$

Approximating ∂B by n chords with nodal points at their centres, we compute the normal components x_i', y_i'; $i = 1, 2, \ldots, n$, by means of formulae (11.1.3), (11.1.4) respectively. Then, since $\phi' = x'$ in equation (13.1.2), from (13.6.1) and (13.6.5) we have immediately the right-hand sides of the approximating equations (13.1.4) which we solve by Method E above. Having thus obtained $\tilde{\sigma}$, we approximate ϕ_x and ϕ_y at the nodal points by

$$\tilde{\phi}_x(\mathbf{q}_i) = \sum_{j=1}^{n} \sigma_j \int_j \frac{\partial}{\partial x} \log|\mathbf{q}_i - \mathbf{q}| \, dq + \pi \sigma_i x_i', \tag{13.6.8}$$

corresponding to (10.5.11), and

$$\tilde{\phi}_y(\mathbf{q}_i) = \sum_{j=1}^{n} \sigma_j \int_j \frac{\partial}{\partial y} \log|\mathbf{q}_i - \mathbf{q}| \, dq + \pi \sigma_i y_i'. \tag{13.6.9}$$

In these formulae we evaluate the coefficients of σ_j by analytic integration along the chords, using results (11.4.4)–(11.4.7). Finally, we approximate C_p above by

$$\tilde{C}_p = 2\tilde{\phi}_x - \tilde{\phi}_x^2 - \tilde{\phi}_y^2 \tag{13.6.10}$$

at each of the nodal points.

Results: Figs. 13.6.1 and 13.6.2 illustrate results obtained for two symmetric aerofoils:

The first, Fig. 13.6.1, is a Joukowski aerofoil obtained from the circle

$$|\zeta| = 0{\cdot}6 \tag{13.6.11}$$

by the transformation

$$z = \zeta - 0{\cdot}1 + 0{\cdot}25/(\zeta - 0{\cdot}1). \tag{13.6.12}$$

For this, there is an analytic solution given by the complex potential

$$w = -(\zeta + 0{\cdot}36/\zeta), \tag{13.6.13}$$

in terms of which

$$|\mathbf{V}| = \left|\frac{dw}{dz}\right| = \left|\frac{dw}{dz}\right| \bigg/ \left|\frac{dz}{d\zeta}\right|, \qquad C_p = 1 - |\mathbf{V}|^2. \qquad (13.6.14)$$

The diagram compares computed values of $\tilde{C}_p$ for $n = 64$, with C_p as given by (13.6.14). The interval points chosen for this computation correspond to equally spaced points on the circle (13.6.11), and they are therefore most closely packed around the leading and trailing edges of the aerofoil.

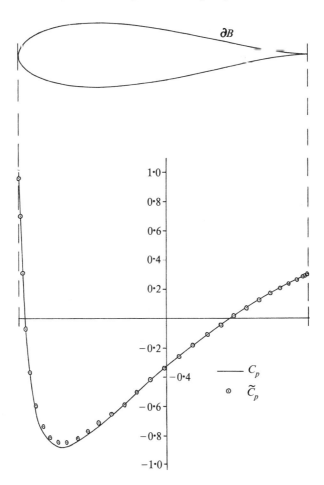

FIG. 13.6.1.

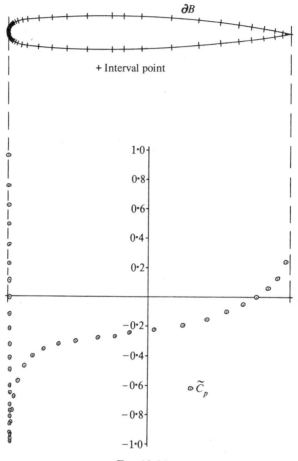

Fig. 13.6.2.

The second aerofoil, Fig. 13.6.2, is a more practical shape whose outline is given by 92 points including the trailing edge. We let these points, which are denoted by crosses in the diagram, be the interval points of ∂B. The corresponding ($n = 92$) nodal values of $\tilde{C}_p$ are illustrated graphically.

Comments: On account of the symmetry of the aerofoils considered here, only $n/2$ equations need to be solved in each case. In Fig. 13.6.2, the sharp negative peak in $\tilde{C}_p$ near the leading edge is due to the rapidly changing curvature of the boundary around that point.

Note that the simple-layer potential method has been extensively applied to fluid flow problems in both two and three dimensions, including flow past lifting aerofoils. For an account see the work of Hess and Smith (1967), which also contains a brief survey of the literature on integral equation methods at that time. More recent developments in this field are described by van de Vooren and Botta (1972) and Hess (1973, 1974, 1975a, b, c).

14

Applications of Green's Boundary Formula

14.1 Introduction and methods

In Section 4.4, we noted the theoretical advantage of formulating the interior Dirichlet problem in terms of Green's boundary formula rather than by means of the simple-layer logarithmic potential. In this chapter, we shall illustrate the practical advantages of Green's formula, which arise

(a) when the values of the harmonic function ϕ and its normal derivative ϕ' on the boundary ∂B are of direct physical significance,

(b) in mixed boundary-value problems, where ϕ is given on part of ∂B and ϕ' (or a linear combination of ϕ and ϕ') on the remainder of the boundary, and

(c) when the boundary ∂B contains corners.

In cases (a) and (b), the advantages of Green's boundary formula, relating as it does the values of ϕ and ϕ', on ∂B, are obvious. In case (c), however, we recall from (4.1.8) that

$$\sigma(\mathbf{q}) = \frac{1}{2\pi}(\phi_i'(\mathbf{q}) + \phi_e'(\mathbf{q})), \qquad \mathbf{q} \in \partial B, \tag{14.1.1}$$

where ϕ_i' and ϕ_e' are the normal derivatives (directed into their respective domains) of the interior and exterior simple-layer logarithmic potentials generated by σ. It follows that

190

(i) σ becomes infinite in magnitude whenever either ϕ_i' or ϕ_e' becomes infinite, e.g. at a corner,

(ii) the asymptotic behaviour of a singular ϕ' is more readily assessed, through formulae such as (12.3.1), than that of the corresponding source density, and

(iii) as observed in Section 12.3, σ may become infinite on ∂B whilst ϕ and the relevant ϕ', i.e. the derivative into the domain B of interest, remain finite.

From Section 4.4, we note that in general, for a plane domain B with boundary ∂B, Green's formula may be written in the form

$$\int_{\partial B} \log|\mathbf{p} - \mathbf{q}| \, \phi'(\mathbf{q}) \, dq - \int_{\partial B} \log|\mathbf{p} - \mathbf{q}|' \, \phi(\mathbf{q}) \, dq = \eta(\mathbf{p}) \, \phi(\mathbf{p}), \quad (14.1.2)$$

with

$$\eta(\mathbf{p}) = \begin{cases} 2\pi, & \mathbf{p} \in B, \\ \Omega(\mathbf{p}), & \mathbf{p} \in \partial B, \\ 0, & \mathbf{p} \notin B + \partial B, \end{cases} \quad (14.1.3)$$

where $\Omega(\mathbf{p})$ is the 'internal' angle at $\mathbf{p}$, i.e. the angle in B between the tangents to ∂B on either side of $\mathbf{p}$. In particular, when ∂B passes smoothly through $\mathbf{p}$, $\Omega(\mathbf{p}) = \pi$ and, from (14.1.2) and (14.1.3), we have

$$\int_{\partial B} \log|\mathbf{p} - \mathbf{q}| \, \phi'(\mathbf{q}) \, dq - \int_{\partial B} \log|\mathbf{p} - \mathbf{q}|' \, \phi(\mathbf{q}) \, dq - \pi\phi(\mathbf{p}) = 0,$$

$$\mathbf{p} \in \partial B, \quad (14.1.4)$$

which is the usual form of Green's boundary formula.

Discretising equation (14.1.4) in the manner of Section 10.3, we obtain

$$\sum_{j=1}^{n} \phi_j' \int_j \log|\mathbf{q}_i - \mathbf{q}| \, dq - \sum_{j=1}^{n} \phi_j \int_j \log|\mathbf{q}_i - \mathbf{q}|' \, dq - \pi\phi_i = 0;$$

$$i = 1, 2, \ldots, n, \quad (14.1.5)$$

which, given ϕ or ϕ' at each nodal point, yields a system of n simultaneous linear equations in n unknowns. Upon solving these equations, we know both ϕ_j and ϕ'_j for each $j = 1, 2, \ldots, n$ and we may then approximate ϕ by

$$\tilde{\phi}(\mathbf{p}) = \frac{1}{\eta(\mathbf{p})} \sum_{j=1}^{n} \left(\phi'_j \int_j \log |\mathbf{p} - \mathbf{q}| \, dq - \phi_j \int_j \log |\mathbf{p} - \mathbf{q}|' \, dq \right),$$

$$\mathbf{p} \in B + \partial B, \qquad (14.1.6)$$

corresponding to (14.1.2).

Whilst the precise equations which we must solve are problem dependent, we again categorise our methods according to the way in which we evaluate the various coefficients—in this case the integrals which appear in formulae (14.1.5) and (14.1.6). Thus, for reference later in this chapter, we define:

Method F: We evaluate the coefficients of ϕ'_j by analytic integration over the two-chord approximation of the interval Δ_j, using formulae (11.3.6), (11.3.15), (11.4.1) and (11.4.2) as in Method A of Chapter 12. The interval length, if required, is approximated by formula (11.1.5) and the coefficients of ϕ_j are evaluated analytically by means of results (11.5.2) and (11.5.3).

Method G: We evaluate the coefficients of ϕ'_j by means of numerical quadrature formula (11.6.2) in conjunction with the analytic result (11.6.3) at the nodal point, as in Method B of Chapter 12 except that we replace the chord lengths in (11.6.3) by the true arc lengths when these are known. Correspondingly we evaluate the interval length accurately, if possible, and the coefficients of ϕ_j are determined analytically as in Method F.

Method H: We evaluate the coefficients of ϕ'_j by means of formulae (11.6.2) and (11.6.3) exactly as in Method B. The coefficients of ϕ_j are evaluated numerically by means of formulae (11.6.1) and (11.6.7), i.e. in a manner analogous to Method C of Chapter 13.

14.2 A heat conduction problem

As a first application of Green's boundary formula (14.1.4), we consider a problem in which both ϕ and ϕ' are of direct physical significance. This is the problem of determining the external thermal resistance—of importance with regard to current rating—of a group of buried electricity cables. Subject to appropriate physical assumptions (Goldenberg, 1969a), the problem may be posed as follows:

Let the cables have uniform cross-section and let them be buried parallel to the plane surface S of the earth. Fig. 14.2.1 shows a typical arrangement, in cross-section, with the x-axis corresponding to the surface S and the contour C denoting the outer boundary of the group of cables. Then the external

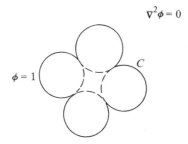

FIG. 14.2.1.

thermal resistance, per unit length of the cable group, is

$$G_C = -g / \int_C \phi'(\mathbf{q}) \, dq, \qquad (14.2.1)$$

where g is the thermal resistivity of the earth and ϕ, denoting temperature, satisfies Laplace's equation in the plane domain illustrated in the Figure, with boundary conditions

$$\left. \begin{array}{l} \phi = 0 \text{ on } S \text{ (and at infinity) and} \\ \phi = 1 \text{ on } C \end{array} \right\}. \qquad (14.2.2)$$

Now, in a problem of this nature, it is preferable for the boundary integral equation approach to remove the infinite boundary. Here, therefore, we eliminate S from the formulation by introducing an image C^* of C with respect to S and setting $\phi = -1$ on C^* as illustrated in Fig. 14.2.2. Then clearly the solution ϕ of Laplace's equation in the infinite domain bounded internally by C and C^*, which tends to zero at infinity and takes the values

$$\left.\begin{array}{l} \phi = 1 \text{ on } C \text{ and} \\ \phi = -1 \text{ on } C^* \end{array}\right\}, \tag{14.2.3}$$

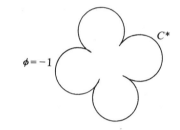

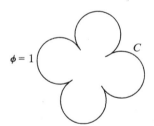

FIG. 14.2.2.

is identical in the lower half-plane to the solution of the original problem (Fig. 14.2.1) with boundary conditions (14.2.2). Thus our problem is to determine G_C from (14.2.1) when ϕ satisfies the Dirichlet boundary conditions (14.2.3).

Letting $\partial B = C + C^*$ and applying Green's boundary formula (14.1.4) to

ϕ, taking into account the boundary conditions, we obtain

$$\int_{\partial B} \log|\mathbf{p} - \mathbf{q}| \, \phi'(\mathbf{q}) \, dq = \pi + \int_C \log|\mathbf{p} - \mathbf{q}|' \, dq - \int_{C^*} \log|\mathbf{p} - \mathbf{q}|' \, dq,$$

$$\mathbf{p} \in C, \qquad (14.2.4)$$

together with a similar equation for $\mathbf{p} \in C^*$. The integrals on the right-hand side may be evaluated analytically by means of results (4.1.23) and (4.1.24). Thus equation (14.2.4) becomes

$$\int_{\partial B} \log|\mathbf{p} - \mathbf{q}| \, \phi'(\mathbf{q}) \, dq = 2\pi, \qquad \mathbf{p} \in C, \qquad (14.2.5)$$

which, in view of the anti-symmetry of ϕ about the x-axis, is sufficient to determine ϕ' on C (or C^*) and hence G_C.

If C bounds a group of m similar cables, we further assume, following Goldenberg (1969b), that the rate of flow of heat is the same from each cable. In this case

$$G = mG_C \qquad (14.2.6)$$

represents the external thermal resistance of each individual cable.

Example 14.2.1

The problem: To determine the external thermal resistance G of each of three similar circular cables of radius a buried in trefoil-touching formation at depth d as illustrated in Fig. 14.2.3.

The method: We apply the formulation above discretised according to Method G of Section 14.1. Thus, dividing the boundary C of the cables (and its image C^* symmetrically) into n equal intervals of arc length h, we approximate equation (14.2.5) by a set of simultaneous linear equations in the approximate values ϕ'_j; $j = 1, 2, \ldots, n$, of ϕ' on C. Correspondingly, we approximate G_C, (14.2.1), by

$$\tilde{G}_C = -g/(h \sum_{j=1}^{n} \phi'_j) \qquad (14.2.7)$$

H

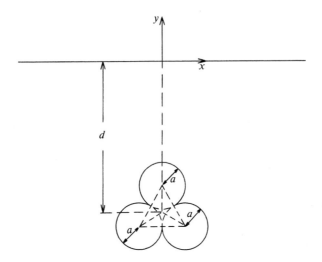

F<small>IG</small>. 14.2.3.

and G, (14.2.6), by

$$\tilde{G} = 3\tilde{G}_C. \tag{14.2.8}$$

On account of the symmetry about the y-axis, we need to solve only $n/2$ equations for an even value of n.

Results: Table 14.2.1 shows computed approximations to G/g for several ratios d/a and three values of n.

Comments: The agreement between the results for successive values of n

T<small>ABLE</small> 14.2.1. Cables in trefoil-touching formation.

	Values of $\tilde{G}/g$		
d/a	$n = 24$	$n = 48$	$n = 96$
5	0·7784	0·7782	0·7781
10	1·1254	1·1251	1·1251
20	1·4598	1·4596	1·4595
40	1·7916	1·7913	1·7913
100	2·2293	2·2291	2·2290

suggests that, for $n = 96$, $\tilde{G}/g$ is correct to at least three decimal places. This accords with results obtained by Goldenberg (1969b) by an analytical method, of more limited application than the present, based upon conformal transformation.

Results similar to those presented here have been obtained for a variety of cable configurations, and for a rectangular trough, by Symm (1969b). Further results, for three cables in close (but not touching) flat formation, have been obtained by Goldenberg (1971) by a generalisation of the present method which allows the cables of a group to lose heat at differing rates and to have different surface temperatures. Solutions of other practical Dirichlet problems are discussed by Christiansen and Rasmussen (1976).

14.3 Torsion problems

To illustrate the application of Green's boundary formula to a Neumann problem, we consider the torsion of a prismatic bar as formulated in terms of the warping function by Jaswon and Ponter (1963).

Note that this problem may also be formulated as a Dirichlet problem in terms of a stress function (Timoshenko and Goodier, 1951) and as such has been treated by integral equation methods based upon the double-layer potential (Mikhlin, 1957; Massonnet, 1965) and the simple-layer potential (Lo and Niedenfuhr, 1970). In the last reference, the warping function is obtained by computing the harmonic conjugate of the simple-layer potential by a method reminiscent of Section 11.7. The Dirichlet formulation also reveals analogies between the torsion problem and problems of membrane deflection, slow viscous flow and, for appropriate geometries, thermal bending of polygonal plates. In the latter context, Green's boundary formula has been applied to the problem by Maiti et al. (1969). In the torsion problem, however, the stress function of the Dirichlet formulation has no direct physical significance whilst the warping function, of the Neumann formulation, is proportional to the displacement parallel to the axis of the cylinder.

Let B denote the uniform cross-section of the prism under consideration. Then the warping function ϕ satisfies Laplace's equation in the domain B with the boundary condition

$$\phi'(\mathbf{q}) = -\frac{\partial}{\partial q}(\tfrac{1}{2}r^2(\mathbf{q})), \qquad \mathbf{q} \in \partial B, \tag{14.3.1}$$

where $r(\mathbf{q})$ is the radial distance of the point $\mathbf{q}$ from some origin in the plane of B. From (14.3.1), it follows immediately that

$$\int_{\partial B} \phi'(\mathbf{q}) \, dq = 0 \qquad (14.3.2)$$

for any boundary ∂B made up of closed contours. Thus, for any bounded domain B, simply- or multiply-connected, the Gauss condition holds and the Neumann problem has a solution which is unique but for an arbitrary additive constant. In terms of this solution, the torsional rigidity of the bar is (Jaswon and Ponter, 1963)

$$G = \iint_B r^2(\mathbf{q}) \, dQ + \int_{\partial B} \phi(\mathbf{q}) \phi'(\mathbf{q}) \, dq \qquad (14.3.3)$$

and the boundary shear stress is

$$S(\mathbf{q}) = \left| \frac{\partial \phi(\mathbf{q})}{\partial q} - (\tfrac{1}{2} r^2(\mathbf{q}))' \right|, \qquad \mathbf{q} \in \partial B, \qquad (14.3.4)$$

ignoring constant factors which depend upon the material concerned. It is evident that G and S are independent of the location of the origin O and of any additive constant attached to the warping function.

Applying Green's boundary formula (14.1.4) to this problem, inserting the boundary condition (14.3.1), we obtain an integral equation for $\phi(\mathbf{q})$, $\mathbf{q} \in \partial B$. We can achieve a unique solution of this equation by prescribing the value of ϕ at one particular point. The solution is obtained, in discrete form, from equations (14.1.5) and we then approximate G by

$$\tilde{G} = \iint_B r^2(\mathbf{q}) \, dQ + \sum_{j=1}^{n} \phi_j \phi'_j h_j \qquad (14.3.5)$$

and S by

$$\tilde{S}(\mathbf{q}) = |\phi^{(1)}(\mathbf{q}) - (\tfrac{1}{2} r^2(\mathbf{q}))'|, \qquad \mathbf{q} \in \partial B, \qquad (14.3.6)$$

where $\phi^{(1)}$ denotes a finite-difference approximation to $\partial \phi / \partial q$.

Example 14.3.1
The problem: To determine the torsional rigidity G and the boundary

shear stress S for a cylindrical bar whose cross-section B is bounded by the ellipse ∂B:

$$\frac{x^2}{a^2} + \frac{y^2}{b^2} = 1. \tag{14.3.7}$$

The analytic solution of this problem is available for comparison (Timoshenko and Goodier, 1951).

The method: Dividing the boundary ∂B into n equal intervals of length h (by means of tables of elliptic integrals), we discretise Green's boundary formula according to Method H of Section 14.1. By taking account of the symmetry of the problem, we reduce the number of linear equations from n to $n/4$ and, at the same time, eliminate the arbitrary constant from the solution by imposing the value $\phi = 0$ on each of the coordinate axes. Solving the equations directly, we evaluate $\tilde{G}$ and $\tilde{S}$ from formulae (14.3.5) and (14.3.6). The integration over B in (14.3.5) is carried out analytically, as is the normal differentiation in (14.3.6). In the latter, a finite-difference approximation of the form

$$\phi^{(1)}(\mathbf{q}_{k+\frac{1}{2}}) = \frac{\phi_{k+1} - \phi_k}{h} \tag{14.3.8}$$

is used to evaluate $\tilde{S}$ at the interval points of ∂B.

Results: Table 14.3.1 illustrates results obtained for an ellipse with axial ratio 2:1. The value of $\tilde{G}$ and the maximum computed value of $\tilde{S}$ are given for successively increasing values of n; the corresponding analytic results are included for comparison.

Comments: The results for $n = 32$, obtained by solving only 8 equations,

TABLE 14.3.1. Ellipse ($a = 2b$).

n	$\tilde{G}/b^4$	$\tilde{S}_{max}/b$
8	4·985	2·134
16	5·087	1·645
32	5·033	1·603
Analytic	5·026	1·600

are accurate to well within 2%. The boundary shear stress, computed or analytic, has its maximum value at the ends of the minor axes. For further details of this and similar examples see Jaswon and Ponter (1963) and, for an extension of the method to elasto-plastic torsion, see Ponter (1966a), Mendelson and Albers (1975).

Example 14.3.2

The problem: To determine the torsional rigidity G and the boundary shear stress S for a hollow square bar whose cross-section B is bounded externally and internally by similar concentric squares ∂B_0 and ∂B_1 with sides of lengths 2a and 2b respectively.

The method: In this example we divide each of the boundaries ∂B_0 and ∂B_1 into $n/2$ equal intervals; consequently the interval length h_0 on ∂B_0 is larger than the interval length h_1 on ∂B_1. Otherwise, we proceed as in the previous example, using symmetry to reduce the number of equations and boundary unknowns to $n/8$.

Results: Table 14.3.2 shows results obtained for $a = 2b$. On the internal boundary ∂B_1, the shear stress S becomes infinite at the corners, which are re-entrant to the domain B, and the computed results reflect this behaviour. Correspondingly, $\tilde{S}_{max}$ in the Table refers only to the outer boundary ∂B_0, where the local maximum occurs at the mid-point of each side.

TABLE 14.3.2. Hollow square ($a = 2b$).

n	$\tilde{G}/a^4$	$\tilde{S}_{max}/a$
16	1·931	1·264
32	2·041	1·309
64	2·050	1·313
128	2·064	1·314

Comments: The torsional rigidity for $n = 128$, obtained by solving only 16 equations, lies within the bounds

$$2\cdot0635 < G/a^4 < 2\cdot0686 \qquad (14.3.9)$$

derived by Synge (1957) by the hypercircle method, and agrees to four figures with the value obtained by Quinlan (1967) by the edge-function method. Although S is not known analytically in this case, $\tilde{S}_{max}$ converges satisfactorily as n is increased. For further details of this and similar examples see again the work of Jaswon and Ponter (1963) and also Kermanidis (1976a, b). For an extension of the method to related problems of inhomogeneous torsion, as for example when ∂B_1 here bounds another material and the bar is a composite rather than hollow, see also Ponter (1966b), Bobrik and Mikhailov (1974).

14.4 Mixed boundary-value problems

Suppose now that ϕ is prescribed on part ∂B_1 of a boundary ∂B and ϕ' on the remainder $\partial B_2 \equiv \partial B - \partial B_1$. In this case the coupled integral equations have a unique solution unless $\partial B = \Gamma$, which case may be avoided by a simple change of scale (Hayes and Kellner, 1972). (Note from (4.5.12), or from Christiansen (1975, 1976), that $\partial B = \Gamma$ when ∂B has unit transfinite diameter.) Then, if $\partial B_1, \partial B_2$ are divided into n_1, n_2 intervals respectively, $n = n_1 + n_2$ in equations (14.1.5) and (14.1.6).

Example 14.4.1
The problem: To determine the solution ϕ of the mixed boundary-value problem displayed in Fig. 14.4.1, where $ABCD$ is a rectangle and $DO = OA = AB$.

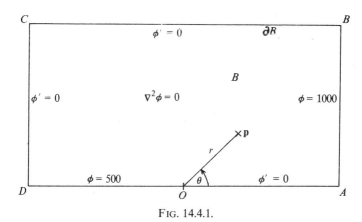

FIG. 14.4.1.

The method: Here ∂B_1 consists of the segments DO and AB of ∂B whilst ∂B_2 is made up of OA and BCD. Dividing ∂B into n equal intervals, where n is a multiple of six, we have $n_2 = 2n_1$ in the notation introduced above. The equations are discretised according to Method F of Section 14.1.

14. Applications of Green's boundary formula

G1	s1	G2	G3	s2	G4	G5	s3	G6	s4	G7	s5
956 955 954	953	955 954 953	954 953 952	952	953 952 951	952 951 950	950	952 951 950		953 951 950	949
911 909 908		910 908 907	908 906 905		906 904 903	904 902 901		903 900 899		903 900 899	
867 864 863	862	865 862 861	862 859 858	857	859 856 854	855 852 850	849	853 849 847		853 848 846	844
824 821 819		821 818 816	817 813 812		811 807 805	806 801 799		801 796 793		800 794 791	
783 779 778	776	779 775 774	773 768 767	765	764 759 757	755 749 746	744	746 740 736		743 736 732	728
744 741 739		739 735 733	730 726 724		718 713 711	703 697 693	670	686 678 674		677 667 662	656
709 705 704	702	703 699 697	692 688 686	684	676 671 669	653 647 644	642	623 612 608	604	578 555 539	500
678 674 673		671 667 665	658 654 653		639 635 634	612 608 606		570 566 564	562	495 499 500	500
651 648 647	645	644 641 639	631 628 626	625	611 608 607	584 581 580	579	546 545 544		498 500 500	500
630 627 626		623 620 618	610 607 606		591 589 587	566 564 563		535 534 534		499 500 500	500
614 611 610	609	606 604 603	594 592 591	590	577 575 574	555 553 552		528 528 527		501 500 500	500
602 600 599		595 593 592	583 582 581		567 566 565	547 546 546		524 524 524		499 500 500	500
595 593 592	591	588 587 586	577 576 575	574	562 561 560	543 543 542	542	522 522 522		500 500 500	500

The numbers in the table have the following significance:

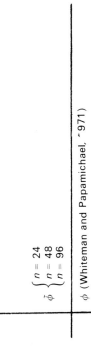

$$\bar{\phi} \begin{cases} n = 24 \\ n = 48 \\ n = 96 \end{cases}$$

ϕ (Whiteman and Papamichael, ˹971)

Fig. 14.4.2

Results: Taking the dimensions of the rectangle to be 14 by 7, we have computed $\tilde{\phi}$ on a unit mesh. In Fig. 14.4.2 we present our results for three values of n, comparing them where possible with a solution due to Whiteman and Papamichael (1971). To the three significant figures given here, the latter solution, obtained by conformal transformation, is essentially equivalent to the true solution of the problem.

Comments: We observe from Fig. 14.4.2 that, as n increases, $\tilde{\phi}$ converges much more slowly near the point O than elsewhere in the rectangle. This is due to the singular nature of the solution at this point. Thus, introducing polar coordinates with origin at O as shown in Fig. 14.4.1, we find following Motz (1946), that at a point $\mathbf{p} = (r, \theta)$ in the neighbourhood of O,

$$\phi(\mathbf{p}) = 500 + \alpha r^{\frac{1}{2}} \cos (\theta/2) + \beta r^{3/2} \cos (3\theta/2) + \dots, \qquad (14.4.1)$$

where $\alpha, \beta, \dots$ are certain constants. It follows that on DO, where $\theta = \pi$,

$$\phi'(\mathbf{p}) = (\alpha/2) r^{-1/2} - (3\beta/2) r^{1/2} + \dots, \qquad (14.4.2)$$

which, if $\alpha \neq 0$, becomes infinite as $r \to 0$. Hence ϕ' has an infinite discontinuity at O and is not well represented by the step-function approximation on DO. Clearly the method must be modified in some way to cope with this situation and one such modification will be described in the next section.

Before considering the treatment of singularities, however, we note that Method F, used in this example, forms the basis of a general purpose computer program (Pitfield and Symm, 1974) for the solution of Laplace's equation in two dimensions. This program, which forms part of the National Physical Laboratory (NPL) Algorithms Library, will solve the equation in an arbitrary domain bounded internally or externally by one or more closed contours, given either ϕ or ϕ' at each point of the boundary. The mathematical formulation and numerical results for a variety of problems are described in an NPL Report (Symm and Pitfield, 1974). The indeterminacy in the solution of the interior Neumann problem is overcome by prescribing a convenient value c for the integral of ϕ around the outermost boundary ∂B_0 of the domain B. Correspondingly, an equation

$$\sum_{j=1}^{n_0} \phi_j h_j = c \qquad (14.4.3)$$

is added to equations (14.1.5), the resulting system being solved by the method of least squares. For this purpose the program, which is available in ALGOL and FORTRAN, uses procedures based upon equations (6.8)–(6.14) of Peters and Wilkinson (1970). Similar programs in FORTRAN have been written by Hayes (1970).

14.5 Treatment of singularities

In Example 14.4.1 we have seen how a boundary singularity in a solution of Laplace's equation adversely affects the rate of convergence of the integral equation method. We shall now show how the method can be modified to overcome this difficulty without appreciably increasing the amount of computation involved. We note, in passing, that other numerical methods, such as those based upon finite differences and finite elements, are no less affected and we refer the reader to the works of Motz (1946), Woods (1953) and Wait and Mitchell (1971), to mention but a few.

Suppose now that around a point $O(r = 0)$ on the boundary ∂B of a domain B the harmonic function ϕ, which satisfies given boundary conditions on ∂B, can be expanded in the series form

$$\phi(\mathbf{p}) = f_0(\mathbf{p}) + \alpha f_1(\mathbf{p}) + \beta f_2(\mathbf{p}) + \gamma f_3(\mathbf{p}) + \ldots, \quad \mathbf{p} = (r, \theta), \quad (14.5.1)$$

where $f_0, f_1, f_2, \ldots$ are known harmonic functions and $\alpha, \beta, \gamma, \ldots$ are unknown constants. Suppose further than f_0 is non-singular in B and that $f_1, f_2, f_3, \ldots$ are arranged in order of ascending powers of r, so that, as $r \to 0$, $f_{i+1}/f_i \to 0$; $i = 1, 2, \ldots$.

If f_1 has a singularity, such as an infinite first derivative, at the point O, we define a function ψ such that

$$\phi(\mathbf{p}) = \psi(\mathbf{p}) + \alpha f_1(\mathbf{p}) + \beta f_2(\mathbf{p}), \quad \mathbf{p} \in B + \partial B. \quad (14.5.2)$$

Then ψ satisfies Laplace's equation in B, with boundary conditions which involve the unknowns α and β, and near the point O:

$$\psi(\mathbf{p}) = f_0(\mathbf{p}) + \gamma f_3(\mathbf{p}) + \ldots. \quad (14.5.3)$$

In this case we apply Green's boundary formula, in its discretised form (14.1.5), to the function ψ in place of ϕ, introducing approximations $\tilde{\alpha}$ and $\tilde{\beta}$

to α and β. Thus, on applying the boundary conditions, we obtain a system of n simultaneous linear algebraic equations in $n + 2$ unknowns, viz. ψ_j or ψ'_j in each of n intervals plus $\tilde{\alpha}$ and $\tilde{\beta}$. We reduce the number of unknowns to n by ignoring f_3 and higher order terms in (14.5.3) and applying this equality, differentiated if necessary, at the two nodal points nearest to O. Solving the resulting n equations directly, we approximate ψ by $\tilde{\psi}$ by analogy with (14.1.6), and hence we obtain the approximation

$$\tilde{\phi}(\mathbf{p}) = \tilde{\psi}(\mathbf{p}) + \tilde{\alpha}f_1(\mathbf{p}) + \tilde{\beta}f_2(\mathbf{p}), \qquad \mathbf{p} \in B + \partial B, \qquad (14.5.4)$$

to the required solution $\phi(\mathbf{p})$.

If f_1 is not singular but f_2 is so, we may proceed in exactly the same way but we may find that we are now unjustified in ignoring f_3 in (14.5.3). In this case, we redefine ψ so that

$$\phi(\mathbf{p}) = \psi(\mathbf{p}) + \alpha f_1(\mathbf{p}) + \beta f_2(\mathbf{p}) + \gamma f_3(\mathbf{p}) + \delta f_4(\mathbf{p}), \qquad \mathbf{p} \in B + \partial B, \qquad (14.5.5)$$

introducing now four extra unknowns α, β, γ and δ into the discretisation of Green's formula in ψ. From (14.5.1) and (14.5.5), it follows that near O

$$\psi(\mathbf{p}) = f_0(\mathbf{p}) + \varepsilon f_5(\mathbf{p}) + \ldots, \qquad (14.5.6)$$

and, ignoring f_5 and higher order terms, we apply this equality, or its normal derivative, at each of the four nodal points nearest to O. We thus again reduce Green's formula in ψ to n equations in n unknowns, from the solution of which $\tilde{\psi}$ follows as before and hence we obtain

$$\tilde{\phi}(\mathbf{p}) = \tilde{\psi}(\mathbf{p}) + \tilde{\alpha}f_1(\mathbf{p}) + \tilde{\beta}f_2(\mathbf{p}) + \tilde{\gamma}f_3(\mathbf{p}) + \tilde{\delta}f_4(\mathbf{p}), \qquad \mathbf{p} \in B + \partial B. \qquad (14.5.7)$$

Each of these methods is illustrated by an example below. Although it is evident that we may similarly treat any number of singularities on the boundary of a finite domain, we restrict our attention here to problems involving only one singularity for the sake of simplicity. For a numerical example involving two singularities, see the work of Symm (1973) from which the following examples are taken. For further examples and a generalisation of the method to cover curved boundaries and discontinuities, as well as singularities, in the boundary conditions, see Papamichael and Symm (1975).

Example 14.5.1

The problem: To determine the solution ϕ of the mixed boundary-value problem displayed in Fig. 14.4.1. (This is the same problem as was considered in Example 14.4.1.)

The method: Recalling the expansion (14.4.1) for ϕ near the point O in Fig. 14.4.1, we observe that ϕ is of the form (14.5.1) with

$$f_0(\mathbf{p}) = 500, \qquad f_1(\mathbf{p}) = r^{1/2} \cos (\theta/2), \qquad f_2(\mathbf{p}) = r^{3/2} \cos (3\theta/2), \qquad \ldots \,;$$

$$\mathbf{p} = (r, \theta). \qquad (14.5.8)$$

It follows, as shown by (14.4.2), that f_1 has a singularity of the type we are currently considering; we therefore define a function ψ, according to (14.5.2), such that

$$\phi(\mathbf{p}) = \psi(\mathbf{p}) + \alpha r^{1/2} \cos (\theta/2) + \beta r^{3/2} \cos (3\theta/2), \qquad \mathbf{p} \in B + \partial B. \qquad (14.5.9)$$

Then ψ is harmonic in B and satisfies on ∂B the boundary conditions

$$\psi'(\mathbf{p}) = 0, \qquad \mathbf{p} \in OA,$$

$$\psi(\mathbf{p}) = 1000 - \alpha r^{1/2} \cos (\theta/2) - \beta r^{3/2} \cos (3\theta/2), \qquad \mathbf{p} \in AB,$$

$$\psi'(\mathbf{p}) - (\alpha/2)r^{-1/2}\sin (\theta/2) - (3\beta/2) r^{1/2}\sin (\theta/2), \qquad \mathbf{p} \in BC, \qquad (14.5.10)$$

$$\psi'(\mathbf{p}) = -(\alpha/2)r^{-1/2} \cos (\theta/2) - (3\beta/2)r^{1/2} \cos (\theta/2), \qquad \mathbf{p} \in CD,$$

$$\psi(\mathbf{p}) = 500, \qquad \mathbf{p} \in DO.$$

We discretise Green's formula in ψ in the manner of Example 14.4.1, introducing $\tilde{\alpha}$ and $\tilde{\beta}$ as above balanced by the approximations

$$\psi_1 = 500, \qquad \psi'_n = 0, \qquad (14.5.11)$$

corresponding to the current form of (14.5.3), with intervals of ∂B numbered anti-clockwise from O. Following the analysis above, we thus obtain

$$\tilde{\phi}(\mathbf{p}) = \tilde{\psi}(\mathbf{p}) + \tilde{\alpha} r^{1/2} \cos (\theta/2) + \tilde{\beta} r^{3/2} \cos (3\theta/2), \qquad \mathbf{p} \in B + \partial B, \qquad (14.5.12)$$

corresponding to (14.5.4).

Results: As in Example 14.4.1, results have been computed for the rectangle with dimensions 14×7. Table 14.5.1 shows how $\tilde{\alpha}$ and $\tilde{\beta}$, which are typical of the solution as a whole, converge as n is repeatedly doubled from 6 up to 96. Fig. 14.5.1 shows the final solution $\tilde{\phi}$ on a square mesh of length 0·25 in the neighbourhood of the singular point O. For comparison, as in Fig. 14.4.2, we include values (essentially those of the true solution) from Whiteman and Papamichael (1971).

Comments: By taking into account the nature of the singularity at O, we have improved greatly upon the results of Example 14.4.1 without appreciably increasing the amount of computation involved. Indeed, it is evident from Table 14.5.1 that an improved solution may thus be obtained even by solving

TABLE 14.5.1. Rectangle (14×7).

n	$\tilde{\alpha}$	$\tilde{\beta}$
6	157·05	3·59
12	152·13	4·61
24	151·75	4·65
48	151·65	4·69
96	151·63	4·71

fewer equations than previously. The results presented here also compare favourably with finite difference and finite element solutions given by Whiteman and Papamichael (1972). Further numerical examples of this nature, including a computation of the capacitance of a rectangular microstrip transmission line, are described elsewhere (Symm, 1973).

Example 14.5.2

The problem: To determine the solution ϕ of the mixed boundary-value problem displayed in Fig. 14.5.2.

The method: In this example the boundary ∂B has a re-entrant corner—a feature which, like the change in the boundary condition in the previous problem, often gives rise to a singularity. Here the corner, which we take to be the origin O of polar coordinates as shown, has an interior angle of $3\pi/2$, whence, from Motz (1946) or more generally Lehman (1959), we have near O

$$\phi(\mathbf{p}) = \alpha + \beta r^{2/3} \cos(2\theta/3) + \gamma r^{4/3} \cos(4\theta/3) + \delta r^2 \cos 2\theta + \ldots,$$

$$\mathbf{p} = (r, \theta), \qquad (14.5.13)$$

Each cell shows two values: the upper value is $\bar\phi\ (n = 96)$ and the lower value is ϕ (Whiteman and Papamichael, 1971).

561·95 / 561·95	569·47 / 569·47	578·77 / 578·77	590·17 / 590·18	603·76 / 603·77	619·18 / 619·19	635·72 / 635·72	652·67 / 652·67	669·54 / 669·54
548·41 / 548·42	555·12 / 555·13	563·95 / 563·96	575·65 / 575·65	590·63 / 590·63	608·30 / 608·30	627·24 / 627·24	646·20 / 646·20	664·59 / 664·59
533·42 / 533·42	538·69 / 538·70	546·24 / 546·24	557·64 / 557·64	574·61 / 574·61	596·23 / 596·23	618·85 / 618·85	640·37 / 640·37	660·40 / 660·40
517·12 / 517·12	520·11 / 520·11	524·81 / 524·81	533·59 / 533·59	553·18 / 553·19	583·67 / 583·67	611·86 / 611·86	636·07 / 636·07	657·52 / 657·52
500·00 / 500·00	500·00 / 500·00	500·00 / 500·00	500·00 / 500·00	500·000 / 500·000	576·41 / 576·41	608·91 / 608·91	634·45 / 634·45	656·48 / 656·48

$\bar\phi\ (n = 96)$

ϕ (Whiteman and Papamichael, 1971)

Fig. 14.5.1

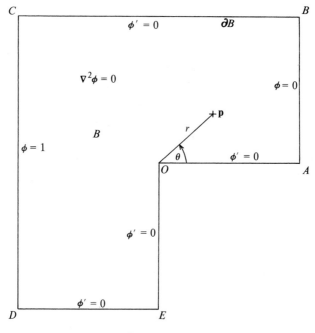

FIG. 14.5.2.

where, $\alpha, \beta, \gamma, \ldots$ are certain constants. This expansion is of the form (14.5.1) with

$$f_0(\mathbf{p}) = 0, \qquad f_1(\mathbf{p}) = 1, \qquad f_2(\mathbf{p}) = r^{2/3} \cos(2\theta/3)\ldots, \quad (14.5.14)$$

in which f_2 has a singularity of the type with which we are concerned. Following (14.5.5) therefore, we define ψ such that

$$\phi(\mathbf{p}) = \psi(\mathbf{p}) + \alpha + \beta r^{2/3} \cos(2\theta/3) + \gamma r^{4/3} \cos(4\theta/3) + \delta r^2 \cos 2\theta, \quad (14.5.15)$$

and proceed to discretise Green's formula in ψ as in the previous example. We again number the intervals of ∂B from the point O, in this case balancing the four extra unknowns $\tilde{\alpha}, \tilde{\beta}, \tilde{\gamma}$ and $\tilde{\delta}$ by the approximations

$$\psi_1 = 0, \qquad \psi_2 = 0, \qquad \psi_{n-1} = 0, \qquad \psi_n = 0, \qquad (14.5.16)$$

which follow from (14.5.6) on ignoring $r^{8/3} \cos(8\theta/3)$ and higher order terms.

0.0000	0.1134	0.2261	0.3371	0.4453	0.5495	0.6487	0.7424	0.8313	0.9166	1.0001
0.0000	0.1140	0.2257	0.3376	0.4452	0.5500	0.6486	0.7426	0.8315	0.9163	1.0000
0.0000	0.1138	0.2269	0.3385	0.4474	0.5521	0.6515	0.7449	0.8831	0.9175	1.0000
0.0000	0.1139	0.2271	0.3387	0.4477	0.5524	0.6516	0.7450	0.8331	0.9175	1.0000
0.0000	0.1147	0.2291	0.3425	0.4536	0.5604	0.6603	0.7527	0.8387	0.9204	1.0000
0.0000	0.1149	0.2294	0.3428	0.4539	0.5606	0.6605	0.7528	0.8388	0.9204	1.0000
0.0000	0.1161	0.2323	0.3486	0.4642	0.5756	0.6772	0.7671	0.8487	0.9254	1.0000
0.0000	0.1162	0.2326	0.3490	0.4646	0.5760	0.6774	0.7672	0.8487	0.9254	1.0000
0.0000	0.1172	0.2352	0.3549	0.4780	0.6019	0.7066	0.7858	0.8633	0.9325	1.0000
0.0000	0.1174	0.2355	0.3555	0.4788	0.6026	0.7066	0.7857	0.8632	0.9324	1.0000
0.0000	0.1177	0.2364	0.3579	0.4869	0.6667	0.7565	0.8210	0.8818	0.9412	1.0000
0.0000	0.1170	0.2371	0.3580	0.4884	0.6663	0.7560	0.8227	0.8816	0.9411	1.0000
					0.7961	0.8154	0.8553	0.9015	0.9503	1.0000
					0.7948	0.8146	0.8548	0.9013	0.9502	1.0000
					0.8487	0.8587	0.8843	0.9191	0.9585	1.0000
					0.8482	0.8581	0.8839	0.9188	0.9584	1.0000
					0.8793	0.8860	0.9048	0.9322	0.9648	1.0000
					0.8787	0.8856	0.9044	0.9319	0.9647	1.0000
					0.9857	0.9012	0.9163	0.9400	0.9687	1.0000
					0.8953	0.9007	0.9163	0.9398	0.9686	1.0000
					0.9009	0.9060	0.9205	0.9427	0.9700	0.9999
					0.9005	0.9055	0.9252	0.9425	0.9698	1.0000

$\hat{\phi}$ ($n = 80$)

ϕ (Papamichael and Whiteman, 1972)

Fɪɢ 14.5.3

We finally obtain

$$\tilde{\phi}(\mathbf{p}) = \tilde{\psi}(\mathbf{p}) + \tilde{\alpha} + \tilde{\beta}r^{2/3}\cos(2\theta/3) + \tilde{\gamma}r^{4/3}\cos(4\theta/3) + \tilde{\delta}r^2\cos 2\theta. \quad (14.5.17)$$

Results: Table 14.5.2 and Fig. 14.5.3 show computed results for the geometry $OA = AB = 5$, $BC = CD = 10$, etc. The Table indicates the measure of agreement between solutions for two successive values of n, whilst the Figure displays on a unit mesh the solution corresponding to the higher value. Also included in the Figure, for comparison, are results obtained by Papamichael and Whiteman (1972) by a numerical conformal transformation method.

TABLE 14.5.2. L-shaped domain.

n	$\tilde{\alpha}$	$\tilde{\beta}$	$\tilde{\gamma}$	$\tilde{\delta}$
40	0·666669	−0·154616	−0·025126	0·000003
80	0·666668	−0·154601	−0·025130	0·000000

Comments: In Figure 14.5.3, the values given for comparison may be no more accurate than those computed by the present method, but the agreement between the two solutions inspires confidence in both. Similar results have been obtained for other geometries too (Symm, 1973).

Finally, in this chapter, we note that the examples which we have described, though illustrative, are by no means exhaustive. Green's boundary formula is a very versatile tool in the solution of Laplace's equation and may be applied successfully to much more complex problems than those considered here. See, for example, Symm (1970) for an application of the method to a problem involving mixed boundary conditions of the linear form (2.1.3) and a boundary containing a cut.

15

Biharmonic Problems

15.1 Introduction

As observed in Chapter 9, many two-dimensional problems of elasto-statics are governed by the biharmonic equation

$$\nabla^4 \chi = 0, \qquad (15.1.1)$$

whose solution χ may be expressed in terms of two harmonic functions ϕ and ψ. Consequently such problems may be treated by the methods of scalar potential theory, and our aim in the present chapter is to demonstrate the practical value of this approach with particular reference to problems of bending and stretching of thin plates as formulated in Sections 9.2 and 9.3.

We concentrate upon the representations (9.1.1) and (9.1.16) for χ. Thus, in a finite simply-connected domain B, bounded by a closed contour ∂B, let

$$\chi = r^2 \phi + \psi; \qquad r^2 = x^2 + y^2, \qquad (15.1.2)$$

where the origin of coordinates $r = 0$ lies inside the domain B. In an infinite domain B, bounded internally by a closed contour ∂B, let

$$\chi = r^2 \phi + \psi + k, \qquad (15.1.3)$$

where k denotes a constant and $r = 0$ lies in the region bounded externally by ∂B. In a multiply-connected domain B, bounded externally by ∂B_0 and internally by $\partial B_1, \partial B_2, \ldots, \partial B_m$, we use the representation (15.1.2), taking the origin to lie within one of the inner boundaries ∂B_j; $1 \leqslant j \leqslant m$.

For each of these representations, the first derivatives of χ with respect to x and y are

$$\left.\begin{array}{l} \chi_x = r^2\phi_x + 2x\phi + \psi_x \\[2mm] \chi_y = r^2\phi_y + 2y\phi + \psi_y \end{array}\right\} \tag{15.1.4}$$

and the second derivatives are

$$\left.\begin{array}{l} \chi_{xx} = r^2\phi_{xx} + 4x\phi_x + 2\phi + \psi_{xx} \\[2mm] \chi_{yy} = r^2\phi_{yy} + 4y\phi_y + 2\phi + \psi_{yy} \\[2mm] \chi_{xy} = r^2\phi_{xy} + 2(y\phi_x + x\phi_y) + \psi_{xy} \end{array}\right\}. \tag{15.1.5}$$

We note from (15.1.4) that the normal derivatives of χ at a point on the boundary

$$\chi' \equiv \chi_x x' + \chi_y y' = r^2\phi' + 2(xx' + yy')\phi + \psi', \tag{15.1.6}$$

and we note from (15.1.5) that

$$\nabla^2\chi \equiv \chi_{xx} + \chi_{yy} = 4(x\phi_x + y\phi_y + \phi) \tag{15.1.7}$$

within the domain B.

Following (9.1.6) and (9.1.7), we express ϕ and ψ as simple-layer potentials

$$\phi(\mathbf{p}) = \int_{\partial B} \log|\mathbf{p} - \mathbf{q}|\, \sigma(\mathbf{q})\, dq, \qquad \mathbf{p} \in B + \partial B, \tag{15.1.8}$$

$$\psi(\mathbf{p}) = \int_{\partial B} \log|\mathbf{p} - \mathbf{q}|\, \zeta(\mathbf{q})\, dq, \qquad \mathbf{p} \in B + \partial B. \tag{15.1.9}$$

Then, approximating ϕ and ψ by $\tilde{\phi}$ and $\tilde{\psi}$ in the manner of Section 10.2, we approximate (15.1.2) by

$$\tilde{\chi}(\mathbf{p}) = r^2(\mathbf{p})\tilde{\phi}(\mathbf{p}) + \tilde{\psi}(\mathbf{p})$$

$$= r^2(\mathbf{p}) \sum_{j=1}^{n} \sigma_j \int_j \log|\mathbf{p} - \mathbf{q}|\,dq + \sum_{j=1}^{n} \zeta_j \int_j \log|\mathbf{p} - \mathbf{q}|\,dq, \qquad (15.1.10)$$

and similarly (15.1.3) by

$$\tilde{\chi}(\mathbf{p}) = r^2(\mathbf{p})\tilde{\phi}(\mathbf{p}) + \tilde{\psi}(\mathbf{p}) + \tilde{k}, \qquad \mathbf{p} \in B + \partial B, \qquad (15.1.11)$$

where $\tilde{k}$ approximates k. In each case approximations to the derivatives of χ, i.e. (15.1.4) and (15.1.5), are typically

$$\tilde{\chi}_x = r^2\tilde{\phi}_x + 2x\tilde{\phi} + \tilde{\psi}_x \qquad (15.1.12)$$

and

$$\tilde{\chi}_{yy} = r^2\tilde{\phi}_{yy} + 4y\tilde{\phi}_y + 2\tilde{\phi} + \tilde{\psi}_{yy}, \qquad (15.1.13)$$

where the derivatives of $\tilde{\phi}$ and $\tilde{\psi}$ are computed as described in Section 10.5. Note that the difficulties associated with the second derivatives of the logarithmic potential on ∂B may be obviated, to a certain extent, by computing the Laplacian of χ, since, from (15.1.7), this may be expressed in terms of ϕ and its first derivatives only.

The above formulation was first applied, to boundary traction problems, by Jaswon et al. (1967), who observed that if the boundary ∂B had corners, the numerically determined source densities $\tilde{\sigma}$ and $\tilde{\zeta}$ tended to oscillate, precluding the computation of an accurate solution near those points. Consequently, such corners were rounded-off, being replaced by circular arcs whose radii decreased as the subdivision of the boundary became finer. In the sections which follow we shall adopt this procedure wherever necessary. We note however that it is possible to give special treatment to corner singularities; cf. Section 14.5 and see also Bernal and Whiteman (1970), Morley (1973).

15.2 Traction problems

We consider first the problem of determining the distribution of stress in a plane domain B which is subject to prescribed (equilibrated) boundary tractions $\mathbf{T}$, with components T_x and T_y per unit length of the boundary ∂B. As in Section 9.3, we express the stress components in terms of the Airy stress function χ, i.e.

$$\Phi_{xx} = \frac{\partial^2 \chi}{\partial y^2}, \qquad \Phi_{yy} = \frac{\partial^2 \chi}{\partial x^2}, \qquad \Phi_{xy} = -\frac{\partial^2 \chi}{\partial x \partial y}; \qquad (15.2.1)$$

then χ satisfies the biharmonic equation (15.1.1) in B with boundary conditions of the form

$$\chi(\mathbf{p}) = f_1(\mathbf{p}), \qquad \chi'(\mathbf{p}) = f_2(\mathbf{p}); \qquad \mathbf{p} \in \partial B. \qquad (15.2.2)$$

In this section we assume that ∂B is a single closed contour, in which case f_1 and f_2 may be derived explicitly from $\mathbf{T}$. The more complex problem of a multiply-connected domain will be discussed in the next section.

Suppose now that B is a finite simply-connected domain. Then, applying the analysis of the previous section to the boundary conditions (15.2.2), we obtain coupled integral equations

$$\left. \begin{aligned} & r^2(\mathbf{p}) \int_{\partial B} \log|\mathbf{p} - \mathbf{q}| \, \sigma(\mathbf{q}) \, dq + \int_{\partial B} \log|\mathbf{p} - \mathbf{q}| \, \zeta(\mathbf{q}) \, dq = f_1(\mathbf{p}), \\[2ex] & r^2(\mathbf{p}) \left\{ \int_{\partial B} \log'|\mathbf{p} - \mathbf{q}| \, \sigma(\mathbf{q}) \, dq + \pi\sigma(\mathbf{p}) \right\} \\[2ex] & \qquad + 2\{x(\mathbf{p})x'(\mathbf{p}) + y(\mathbf{p})y'(\mathbf{p})\} \int_{\partial B} \log|\mathbf{p} - \mathbf{q}| \, \sigma(\mathbf{q}) \, dq \\[2ex] & \qquad + \int_{\partial B} \log'|\mathbf{p} - \mathbf{q}| \, \zeta(\mathbf{q}) \, dq + \pi\zeta(\mathbf{p}) = f_2(\mathbf{p}), \end{aligned} \right\} \mathbf{p} \in \partial B, \ (15.2.3)$$

for the source densities σ and ζ. Dividing ∂B into n intervals and discretising equations (15.2.3) at one nodal point in each, we obtain a system of $2n$ linear equations in the $2n$ unknowns $\sigma_1, \sigma_2, \ldots, \sigma_n; \zeta_1, \zeta_2, \ldots, \zeta_n$. Solving these equations directly, we compute $\tilde{\chi}$ from (15.1.10) and its derivatives as required.

In the case of an exterior domain B, in which we adopt the representation (15.1.3) for χ, the first boundary condition (15.2.2) yields the integral equation

$$r^2(\mathbf{p}) \int_{\partial B} \log |\mathbf{p} - \mathbf{q}| \, \sigma(\mathbf{q}) \, dq + \int_{\partial B} \log |\mathbf{p} - \mathbf{q}| \, \zeta(\mathbf{q}) \, dq + k = f_1(\mathbf{p}), \quad (15.2.4)$$

in place of the first of equations (15.2.3). The second equation is unchanged, and the extra unknown k is balanced by the extra condition

$$\int_{\partial B} \sigma(\mathbf{q}) \, dq = 0 \qquad (15.2.5)$$

which ensures the uniqueness of χ (see Section 9.1). From the solution of $2n + 1$ linear equations, we compute $\tilde{\chi}$ from (15.1.11) and its derivatives similarly.

In the examples which follow, in both this section and the next, we compute in particular such combinations $\tilde{S}$ and $\tilde{H}$ of the derivatives of $\tilde{\chi}$ as approximate the maximum shear stress

$$S = 0.5 \left\{ (\Phi_{xx} - \Phi_{yy})^2 + 4\Phi_{xy}^2 \right\}^{\frac{1}{2}} \qquad (15.2.6)$$

and the hydrostatic pressure or total stress

$$H = \Phi_{xx} + \Phi_{yy}. \qquad (15.2.7)$$

As in previous chapters, we categorise our precise methods of computation according to the ways in which we evaluate the various coefficients of σ_j and ζ_j in the discretisation; thus:

Method I: We approximate the integral of the logarithmic kernel by Simpson's rule (11.6.2) in conjunction with the analytic result (11.6.3) at the nodal point. We approximate the integrals of derivatives of this kernel by the mid-ordinate rule (11.6.1) in conjunction with result (11.6.5) at the nodal point. In each case we use exact interval length rather than chord length.

Method J: We evaluate the coefficients as in Method I except that we use result (11.6.8) in place of (11.6.5).

Example 15.2.1

The problem: To determine the stress distribution in a rectangle of dimensions $2a \times 2b$ ($a > b$) with semicircular portions of radius c removed from the middle of each of the longer sides and uniform tension, T per unit length, applied to each of the shorter sides (Fig. 15.2.1).

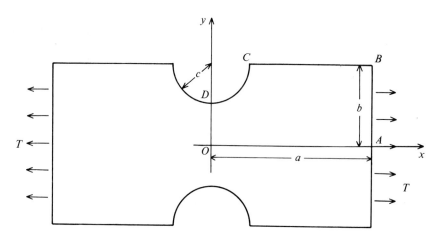

Fɪɢ. 15.2.1.

The method: The only non-zero boundary tractions in this problem, for coordinate axes as illustrated, are $T_x = \pm T$ on the ends $x = \pm a$ of the rectangle. It is therefore a simple matter to express the boundary conditions in terms of the Airy stress function χ by the analysis described in Section 9.3. Thus, choosing the constants of integration so as to preserve the symmetry of the problem we obtain on the quadrant $ABCD$ of ∂B the boundary values

$$\chi = \begin{cases} 0.5\, Ty^2 \text{ on } AB, \\ 0.5\, Tb^2 \text{ on } BC, \\ Tby - 0.5\, Tb^2 \text{ on } CD, \end{cases} \tag{15.2.8}$$

$$\chi' = \begin{cases} 0 \text{ on } AB, \\ -Tb \text{ on } BC, \\ Tby' \text{ on } CD. \end{cases} \tag{15.2.9}$$

In practice these boundary conditions are modified slightly since we round-off the corners at the points B and C, but essentially formulae (15.2.8) and (15.2.9) define the functions f_1 and f_2 in equations (15.2.3). We discretise this problem by Method I above, utilising symmetry to reduce the integral equations to $n/2$ simultaneous linear equations which we solve directly.

Results: This problem has been solved for the case $a = 2\cdot0$, $b = 0\cdot5$, $c = 0\cdot25$ and $T = 1\cdot0$, using two successive subdivisions of the boundary with corners rounded-off as shown in Fig. 15.2.2. From values of the stress com-

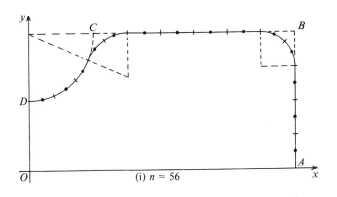

(i) $n = 56$

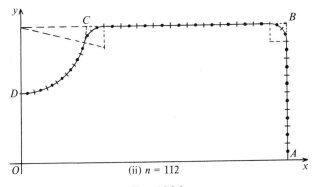

(ii) $n = 112$

Fig. 15.2.2.

ponents computed on a square mesh of length 0·0625 for $n = 112$, contours of constant maximum shear stress and constant hydrostatic tension have been drawn in Fig. 15.2.3.

Comments: Whilst solving a practical problem of this nature for two, or more, values of n provides a useful consistency check on the results obtained, we also find it advisable to test our numerical procedures on a trial problem, with known solution, in the given domain. In the present case such a problem is provided most appropriately by the trial function $\chi = y^2$, which would be proportional to the true stress function in the rectangle if there were no cutouts. For further details of this and related problems see Jaswon *et al.* (1967).

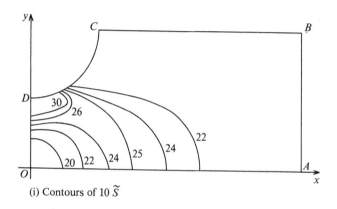

(i) Contours of $10\,\tilde{S}$

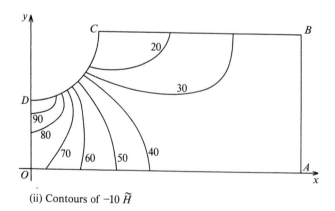

(ii) Contours of $-10\,\tilde{H}$

Fig. 15.2.3.

Example 15.2.2

The problem: To determine the stress distribution in an infinite medium around an elliptic cavity ∂B:

$$\frac{x^2}{a^2} + \frac{y^2}{b^2} = 1; \qquad a \geqslant b, \tag{15.2.10}$$

subject to uniform normal internal pressure P per unit length as illustrated in Fig. 15.2.4.

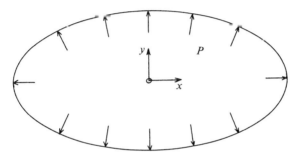

FIG. 15.2.4.

This problem has an analytic solution (Bhargava, 1959) from which, in particular, the maximum shear stress at a point (x, y) is given by

$$S(x, y) = P(b^4 x^2 + a^4 y^2)^{1/2} / \{(x^2 + y^2 - c^2)^2 + 4c^2 y^2\}^{3/4}, \tag{15.2.11}$$

where

$$c^2 = a^2 - b^2. \tag{15.2.12}$$

The method: Following the analysis of Section 9.3, choosing the constants of integration so as to preserve maximum symmetry, we obtain in this case the boundary conditions

$$\chi = -P(x^2 + y^2)/2, \tag{15.2.13}$$

$$\chi' = -P(xx' + yy'). \tag{15.2.14}$$

Since χ is an exterior biharmonic, we apply the representation (15.1.3) so that (15.2.13) defines f_1 in equation (15.2.4). The function f_2 in the second of equations (15.2.3) is defined by formula (15.2.14) and the problem is discretised according to Method J above. Taking symmetry into account, the integral equations coupled with the side condition (15.2.5) reduce to $n/2 + 1$ linear algebraic equations.

Results: For the case $a = 2{\cdot}0$, $b = 1{\cdot}0$ and $P = 1{\cdot}0$ we have computed the maximum shear stress at points of a square mesh (of length $0{\cdot}25$) around the cavity. Resulting values of $4000\,\tilde{S}^2$ for $n = 64$ (equal intervals) are compared with corresponding analytic values of $4000\,S^2$ in Fig. 15.2.5.

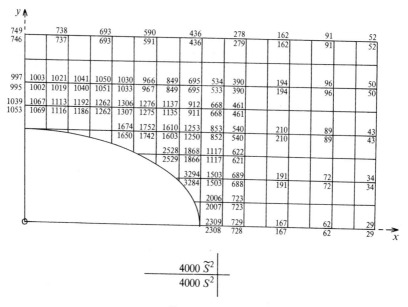

$$\frac{4000\,\tilde{S}^2}{4000\,S^2}$$

Fig. 15.2.5.

Comments: It will be observed in Fig. 15.2.5 that no comparisons are made at mesh points very close to the boundary ∂B. This is because $\tilde{\sigma}$ and $\tilde{\zeta}$, the step-function approximations to σ and ζ for $n = 64$, are too crude to generate $\tilde{S}$ and $\tilde{\chi}$ accurately at these points. Improved results may be obtained (Symm, 1964) by interpolating between values of $\tilde{\chi}$ at some distance from the boundary and known values of χ on ∂B, and by differentiating the interpolant either numerically or analytically. Alternatively, the accuracy of $\tilde{\chi}$ may be increased by evaluating the coefficients of σ_j and ζ_j more accurately. However the improvement in the stress components, for a particular value of n, is limited by the singular nature of the second derivatives of the logarithmic kernel.

It is interesting to note that, at points on the boundary ∂B, the stress components (in any traction problem) may be obtained without computing the second derivatives of χ directly. The given traction components are related to the stresses by

$$-T_x = \Phi_{xx}x' + \Phi_{xy}y' \qquad (15.2.15)$$

$$-T_y = \Phi_{yy}y' + \Phi_{xy}x', \qquad (15.2.16)$$

from which, recalling (15.2.7),

$$\Phi_{xy} = -(Hx'y' + T_xy' + T_yx'). \qquad (15.2.17)$$

Thus, by computing H on ∂B, we may obtain Φ_{xy} and hence Φ_{xx}, Φ_{yy} from (15.2.15), (15.2.16) respectively. But, bearing in mind (15.2.1), we have

$$H = -\nabla^2\chi \qquad (15.2.18)$$

and this, as previously observed, involves ϕ and its first derivatives only. This method has been used by Rim and Henry (1967) to obtain the stress components on the elliptic boundary (15.2.10) of the problem above for several axial ratios including the one considered here.

Ideally, of course, one would prefer to compute the stress components on ∂B, as in B, directly from the approximate stress functions $\tilde{\chi}$ or, more precisely, from the source densities $\tilde{\sigma}, \tilde{\zeta}$. One might then estimate the error in the solution as a whole by checking the compatibility of the computed results with the given boundary tractions in equations (15.2.15) and (15.2.16). Unfortunately the source densities $\tilde{\sigma}, \tilde{\zeta}$ do not have the continuity necessary to ensure the existence of the second derivatives of $\tilde{\phi}, \tilde{\psi}$ on the boundary. Consequently Rim and Henry (1968) have replaced the discrete source densities σ_j, ζ_j by piecewise quadratic functions with $C^{(1)}$ continuity on ∂B. They have thereby improved upon their earlier results for the elliptic cavity problem above.

15.3 Multiply-connected domains

Suppose now that B is a multiply-connected domain, bounded externally by ∂B_0 and internally by $\partial B_1, \partial B_2, \ldots, \partial B_m$, with each boundary contour subject to equilibrated tractions. Then, following the analysis of Section 9.3, the stress function χ satisfies boundary conditions of the form

$$\chi(\mathbf{p}) = f_1(\mathbf{p}), \qquad \chi'(\mathbf{p}) = f_2(\mathbf{p}); \qquad \mathbf{p} \in \partial B_0, \qquad (15.3.1)$$

$$\left. \begin{array}{l} \chi(\mathbf{p}) = f_1(\mathbf{p}) + \alpha_j x(\mathbf{p}) + \beta_j y(\mathbf{p}) + \gamma_j \\[2mm] \chi'(\mathbf{p}) = f_2(\mathbf{p}) + \alpha_j x'(\mathbf{p}) + \beta_j y'(\mathbf{p}) \end{array} \right\} \mathbf{p} \in \partial B_j; \qquad j = 1, 2, \ldots, m, \qquad (15.3.2)$$

where $\alpha_j, \beta_j, \gamma_j; j = 1, 2, \ldots, m$, are constants to be determined. Introducing the representation (15.1.2) for χ, with the origin $r = 0$ enclosed by one of the inner boundaries ∂B_j, we obtain integral equations similar to (15.2.3) for the source densities σ and ζ. Here, however, the integral equations also involve the $3m$ unknowns $\alpha_j, \beta_j, \gamma_j; j = 1, 2, \ldots, m$, and these are balanced by the extra conditions

$$\left. \begin{array}{l} \displaystyle\int_{\partial B_j} x(\mathbf{q})\sigma(\mathbf{q})\,dq = 0 \\[5mm] \displaystyle\int_{\partial B_j} y(\mathbf{q})\sigma(\mathbf{q})\,dq = 0 \\[5mm] \displaystyle\int_{\partial B_j} \sigma(\mathbf{q})\,dq = 0 \end{array} \right\} j = 1, 2, \ldots, m, \qquad (15.3.3)$$

which correspond to (9.3.8).

Dividing $\partial B = \partial B_0 + \partial B_1 + \partial B_2 + \ldots + \partial B_m$ into n intervals and thus discretising equations (15.3.1)–(15.3.3), we obtain $2n + 3m$ linear equations in an equal number of unknowns. From the solution of these equations we may compute $\tilde{\chi}$ and its derivatives as previously. Few problems have yet been solved by means of this formulation but, in view of the difficulties associated with the boundary, it seems expedient to evaluate the coefficients of σ_j and ζ_j as accurately as possible. In particular we suggest the following:

Method K : We approximate (if necessary) each interval of ∂B by the two chords which join its end-points to the nodal point within it (Fig. 11.1.1). We then integrate the logarithmic kernel analytically over each interval using results (11.3.6) and (11.3.15) in general, together with (11.6.3) at the nodal points. Similarly we integrate the partial derivatives of this kernel analytically using results (11.4.4), (11.4.5), (11.3.11)–(11.3.13) and (11.3.16)–(11.3.20) in general. The integral of the normal derivative is obtained by combining the integrals of the first partial derivatives with respect to x and y, following formula (11.2.4), except over the interval containing the nodal point. Here the curvature of the true boundary is taken into account by using formula (11.6.5) or its equivalent (11.5.3).

Example 15.3.1

The problem: To determine the stress distribution in a rectangle with a circular hole in it, as illustrated in Fig. 15.3.1, when

(i) there is uniform normal pressure P per unit length over both the inner and outer boundaries and

(ii) there is uniform normal pressure P per unit length over the inner boundary whilst the outer boundary is traction-free.

Case (i), which is a preliminary test problem for case (ii), has the obvious analytic solution

$$\Phi_{xx} = \Phi_{yy} = -P, \qquad \Phi_{xy} = 0 \qquad (15.3.4)$$

everywhere in the domain B.

The method: Here $m = 1$ in the analysis above and we may therefore drop the subscript j from the constants α, β, γ in equations (15.3.2). In these equations we set

$$f_1(\mathbf{p}) = 0, \qquad f_2(\mathbf{p}) = -Pc; \qquad \mathbf{p} \in \partial B_1 \qquad (15.3.5)$$

corresponding to the given boundary tractions. Similarly, in equations (15.3.1), we set

$$f_1(\mathbf{p}) = -0 \cdot 5P(x^2 + y^2), \qquad f_2(\mathbf{p}) = -P(xx' + yy'); \qquad \mathbf{p} = (x, y) \in \partial B_0$$

$$(15.3.6)$$

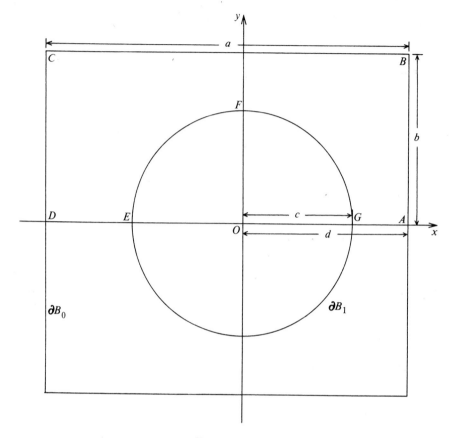

FIG. 15.3.1.

in case (i) and

$$f_1(\mathbf{p}) = 0, \qquad f_2(\mathbf{p}) = 0; \qquad \mathbf{p} \in \partial B_0 \qquad (15.3.7)$$

in case (ii). In each case, the problem is discretised by Method K.

Results: In case (i), the analytic solution corresponding to boundary conditions (15.3.5) and (15.3.6) is

$$\chi = -0{\cdot}5P(x^2 + y^2), \qquad \alpha = 0, \qquad \beta = 0, \qquad \gamma = -0{\cdot}5Pc^2. \quad (15.3.8)$$

Table 15.3.1 compares computed values $\tilde{\alpha}$ and $\tilde{\gamma}$, for various subdivisions

TABLE 15.3.1

n	$\tilde{\alpha}$	$\tilde{\gamma}$
16	0·00714	−71·40
32	0·00036	−72·01
64	0·00016	−72·00
80	0·00001	−72·00
Anal.	0·00000	−72·00

of the boundary, with their analytic counterparts α and γ for data

$$P = 1\cdot0, \quad a = 33\cdot0, \quad b = 15\cdot0, \quad c = 12\cdot0, \quad d = 15\cdot0. \quad (15.3.9)$$

Here $\tilde{\beta} = 0$ from the symmetry of the problem and, if we take this symmetry into account, the second of equations (15.3.3) is automatically satisfied. We need therefore solve only $n + 2$ simultaneous linear equations, where $n = 2(na + 2nb + nc)$ and na, nb, nc denote the numbers of intervals on BC, $AB(CD)$, EFG respectively in Fig. 15.3.1. The values of n in Table 15.3.1

TABLE 15.3.2

Point	$\tilde{\Phi}_{xx}$	$\tilde{\Phi}_{yy}$	$\tilde{\Phi}_{xy}$	$\tilde{S}$
1	−1·000	−1·000	0·000	0·000
2	−1·000	−1·000	0·000	0·000
3	−1·001	−0·999	0·001	0·001
4	−1·000	−0·999	0·003	0·003
5	−0·998	−0·998	0·043	0·043
6	−1·000	−1·000	0·000	0·000
7	−1·001	−1·000	−0·001	0·001
8	−1·001	−0·998	−0·003	0·004
9	−0·997	−0·996	−0·025	0·025
10	−0·997	−1·004	0·000	0·004
11	−0·994	−1·012	0·009	0·013
12	−1·002	−0·993	0·001	0·004
13	−0·996	−1·002	−0·001	0·003
14	−1·003	−0·996	0·000	0·003
15	−0·997	−1·007	−0·005	0·007
16	−1·001	−0·998	0·003	0·003
17	−0·997	−0·997	0·016	0·016
18	−0·999	−0·999	−0·001	0·001
19	−1·002	−0·999	−0·001	0·001
20	−0·999	−1·000	−0·002	0·002
21	−0·995	−1·002	−0·032	0·032

1

correspond to equal intervals on each section of the boundary with $na = nb$ $= nc = 2$, 4 and 8 plus a final subdivision $na = nc = 8$, $nb = 12$. For the latter we also give, in Table 15.3.2, computed stresses $\tilde{\Phi}_{xx}$, $\tilde{\Phi}_{yy}$, $\tilde{\Phi}_{xy}$ and $\tilde{S}$ at points of a square mesh of side 3·0, numbered as indicated in Fig. 15.3.2.

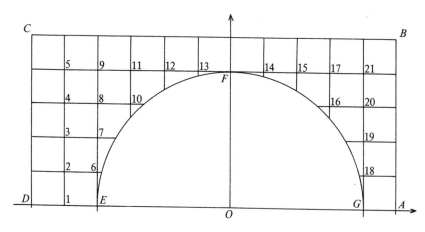

FIG. 15.3.2.

Similar results for case (ii), in which the analytic solution is not known, are given in Tables 15.3.3 and 15.3.4.

Comments: On comparing the results in Table 15.3.2 with the analytic solution (15.3.4) when $P = 1·0$, it is evident that the maximum errors occur near the corners of the domain. For greater accuracy, these corners may be rounded-off as in Example 15.2.1.

TABLE 15.3.3

n	$\tilde{\alpha}$	$\tilde{\gamma}$
16	−0·3345	32·11
32	−0·2509	26·34
64	−0·2578	23·30
80	−0·2569	23·30

TABLE 15.3.4

Point	$\tilde{\Phi}_{xx}$	$\tilde{\Phi}_{yy}$	$\tilde{\Phi}_{xy}$	$\tilde{S}$
1	−0·249	1·836	0·000	1·042
2	−0·108	1·563	0·392	0·923
3	0·032	1·053	0·482	0·702
4	0·160	0·490	0·297	0·340
5	0·113	0·127	−0·151	0·151
6	−0·020	2·704	1·414	1·963
7	−0·085	2·189	1·571	1·939
8	0·585	0·984	0·923	0·944
9	0·457	0·316	0·532	0·537
10	−0·596	3·001	2·441	3·033
11	1·249	0·351	1·076	1·166
12	1·655	0·836	1·075	1·150
13	1·293	−0·428	1·306	1·564
14	2·202	−0·795	−1·299	1·983
15	1·796	1·098	−1·262	1·310
16	0·331	2·844	−2·841	3·107
17	1·239	0·406	−1·050	1·130
18	−0·292	3·019	−1·467	2·212
19	0·406	2·442	−1·679	1·964
20	0·480	1·225	−0·896	0·970
21	0·291	0·359	−0·092	0·098

15.4 Plate bending problems

We consider now the problem of determining the transverse deflection w of a thin plate subject to normal loading T per unit area. Following the analysis of Section 9.2, we express w in the form

$$w = W + \chi, \tag{15.4.1}$$

where W is a particular solution of the equation

$$\nabla^4 W = T/D. \tag{15.4.2}$$

Then χ satisfies the biharmonic equation (15.1.1) in the domain B of the plate, with boundary conditions

$$\chi(\mathbf{p}) = -W(\mathbf{p}), \qquad \mathbf{p} \in \partial B \tag{15.4.3}$$

and, if ∂B is clamped,

$$\chi'(\mathbf{p}) = -W'(\mathbf{p}), \qquad \mathbf{p} \in \partial B \tag{15.4.4}$$

or, if ∂B is simply-supported,

$$\nabla^2\chi(\mathbf{p}) + \frac{1-\nu}{\rho(\mathbf{p})}\chi'(\mathbf{p}) = -\nabla^2 W(\mathbf{p}) - \frac{1-\nu}{\rho(\mathbf{p})}W'(\mathbf{p}), \qquad \mathbf{p} \in \partial B, \tag{15.4.5}$$

where $\rho(\mathbf{p})$ is the 'internal' curvature of ∂B at $\mathbf{p}$.

Applying the analysis of Section 15.1 to this problem, the boundary conditions (15.4.3) and either (15.4.4) or (15.4.5) yield coupled integral equations for the source densities σ and ζ. In the case when ∂B is clamped, these equations are of the same form (15.2.3) as arise in the traction problem. When ∂B is simply-supported, the second integral equation is of a more complex form, but it still involves no higher than first derivatives of ϕ and ψ. Discretising the integral equations as previously, and solving the resulting simultaneous linear equations directly, we compute $\tilde{\chi}$ from (15.1.10). Thus, knowing W, from (9.2.3) or (9.2.4) for example, we obtain the approximation

$$\tilde{w} = W + \tilde{\chi} \tag{15.4.6}$$

to w. Similarly we may approximate the derivatives of w and thus obtain approximations $\tilde{M}_{xx}$, $\tilde{M}_{yy}$ and $\tilde{M}_{xy}$ to the bending and twisting moments

$$M_{xx} = -D\left[\frac{\partial^2 w}{\partial x^2} + \nu\frac{\partial^2 w}{\partial y^2}\right], \tag{15.4.7}$$

$$M_{yy} = -D\left[\frac{\partial^2 w}{\partial y^2} + \nu\frac{\partial^2 w}{\partial x^2}\right], \tag{15.4.8}$$

$$M_{xy} = -M_{yx} = D(1-\nu)\frac{\partial^2 w}{\partial x\partial y}. \tag{15.4.9}$$

Example 15.4.1

The problem: To determine the deflection and the bending moments in a simply-supported square plate, of unit side and Poisson's ratio $\nu = 0\cdot3$, subject to uniform normal loading $T = k$ (constant) per unit area.

This problem has been solved analytically by Timoshenko and Woinowsky-

Krieger (1959) using the reduced boundary condition

$$\nabla^2 \chi(\mathbf{p}) = -\nabla^2 W(\mathbf{p}), \qquad \mathbf{p} \in \partial B, \tag{15.4.10}$$

obtained by putting $1/\rho = 0$ in (15.4.5).

The method: From (9.2.3), we choose the particular integral

$$W = \frac{k}{48D}(x^4 + y^4) \tag{15.4.11}$$

with the origin of coordinates at the centre of the square and axes parallel to the sides. Then, rounding-off the corners as in Example 15.2.1, we discretise the problem according to Method I of Section 15.2, formulating the equations as above from boundary conditions (15.4.3) and either (a) (15.4.5) or (b) (15.4.10).

Results: Table 15.4.1 shows computed values $\partial \tilde{w}/\partial n$ and $\rho^{-1}\partial \tilde{w}/\partial n$ at the nodal points on the corner arcs, for three successive subdivisions of the boundary, in case (a). The basic interval length h on the straight sides equals the radius of curvature on the corners.

TABLE 15.4.1

h	$\partial \tilde{w}/\partial n$	$h^{-1}\partial \tilde{w}/\partial n$
0·2500	0·0095k/D	0·0381k/D
0·1250	0·0053k/D	0·0427k/D
0·0625	0·0028k/D	0·0452k/D

Table 15.4.2 shows $\tilde{w}$, $\tilde{M}_{xx}$ and $\tilde{M}_{yy}$, at selected points on the axes of symmetry of the domain, as computed in cases (a) and (b) for the final subdivision $n = 64$. Also included, and labelled (c), are the corresponding values of w, M_{xx} and M_{yy} obtained by Timoshenko and Woinowsky-Krieger (1959).

Comments: We see from Table 15.4.1 that $\partial \tilde{w}/\partial n$ decreases as $h(=\rho)$ decreases, thereby supporting the conjecture (Timoshenko and Woinowsky-Krieger, 1959) that $\partial w/\partial n = 0$ at a corner. We also observe that $h^{-1}\partial \tilde{w}/\partial n$ increases as h decreases, which implies from $M_{nn} = 0$ that $\partial^2 w/\partial n^2$ increases

TABLE 15.4.2

| x | 0·0 | 0·2 | 0·4 | 0·2 | 0·4 |
y	0·0	0·0	0·0	0·2	0·4
$D\tilde{w}/k$ (a)	0·0042	0·0035	0·0014	0·0029	0·0005
(b)	0·0042	0·0034	0·0014	0·0028	0·0005
(c)	0·0041	—	—	—	—
$\tilde{M}_{xx}/k$ (a)	0·0491	0·0438	0·0220	0·0370	0·0146
(b)	0·0487	0·0433	0·0216	0·0365	0·0114
(c)	0·0479	0·0424	0·0209	—	—
$\tilde{M}_{yy}/k$ (a)	0·0491	0·0409	0·0170	0·0370	0·0146
(b)	0·0487	0·0406	0·0170	0·0365	0·0114
(c)	0·0479	0·0400	0·0168	—	—

as ρ decreases, i.e. that the deflected surface has an appreciable curvature in the n-direction at a corner point in contrast with its zero curvature at other edge points. This supports the physically based view that simply-supported plates have a tendency to ride up at the corners, so requiring reactions of opposite sign to those elsewhere along the edge in order to maintain $w = 0$. The results in Table 15.4.2 suggest that the reduced boundary condition (15.4.10) may be applied to a polygonal domain even when the corners are rounded-off.

For further details of this and other plate bending problems similarly treated see Jaswon and Maiti (1968), Hansen (1976). For applications of the Chakrabarty representation (9.1.21) to simply-supported polygonal plates see Maiti and Chakrabarty (1974). For a problem of a clamped plate subject to a concentrated transverse load see Brown and E. Jaswon (1971), whose results contradict Hadamard's conjecture and agree with the theory of Duffin (1974).

16

Three-Dimensional
Numerical Analysis

16.1 Surface subdivision

As in two dimensions, we wish to subdivide each boundary surface ∂B into elements Δ_j in a manner appropriate to the basic approximation of Section 10.2 for a continuous function u, viz.

$$u(\mathbf{q}) = u_j (\text{constant}) \text{ for } \mathbf{q} \in \Delta_j. \qquad (16.1.1)$$

Thus, changes in the boundary conditions, edges and corners of ∂B must all coincide with edges and corners of individual elements. However, whereas in two dimensions each boundary contour is simply subdivided into intervals, possibly varying in length, in three dimensions a surface may be divided into elements which vary in shape as well as in size. In practice, leaving aside (until Section 16.5) the special class of problems with rotational symmetry, we approximate the surface ∂B by plane polygonal elements with nodal points at their centroids. Any symmetry, including rotational symmetry, is of course taken into account.

16.2 Integration formulae

Suppose that we have approximated a boundary surface ∂B by plane polygonal elements $\Delta_j; j = 1, 2, \ldots, n$. Then in order to apply the integral equation formulations of Part I we must integrate, with respect to $\mathbf{q}$ over each element Δ_j, the kernels (2.2.1), i.e.

$$k(\mathbf{p}, \mathbf{q}) = g(\mathbf{p}, \mathbf{q}), \qquad g'(\mathbf{p}, \mathbf{q}), \qquad g(\mathbf{p}, \mathbf{q})', \tag{16.2.1}$$

where, from (1.1.1),

$$g(\mathbf{p}, \mathbf{q}) = |\mathbf{p} - \mathbf{q}|^{-1}. \tag{16.2.2}$$

Here we follow the example of Laskar (1971).

In general, when $\mathbf{p} \notin \Delta_j$, we integrate numerically by means of the centroid rule

$$\int_j k(\mathbf{p}, \mathbf{q}) \, dq = k(\mathbf{p}, \mathbf{q}_j) A_j, \tag{16.2.3}$$

where $\mathbf{q}_j$ denotes the centroid and A_j the area of the polygon Δ_j. This corresponds to using the mid-ordinate rule (11.6.1) in two dimensions.

When $\mathbf{p} \in \Delta_j$, the kernels (16.2.1) may be integrated analytically over the polygon. In particular, it is immediately evident that

$$\int_j g'(\mathbf{p}, \mathbf{q}) \, dq = 0 = \int_j g(\mathbf{p}, \mathbf{q})' \, dq, \qquad \mathbf{p} \in \Delta_j, \tag{16.2.4}$$

corresponding to result (11.4.8) for a straight line. The analytic formula for the integration of $g(\mathbf{p}, \mathbf{q})$ over the polygon may be derived from the result for a triangle with vertex at $\mathbf{p}$. Thus, if Δ is such a triangle, of area A and with side of lengths a, b and c, then

$$\int_\Delta g(\mathbf{p}, \mathbf{q}) \, dq = \frac{2A}{c} \log\left(\frac{a + b + c}{a + b - c}\right), \tag{16.2.5}$$

where it is assumed that the side of length c is opposite the vertex $\mathbf{p}$. It follows, for example, that if $\mathbf{p}$ is at the centre of a square element Δ_j of side h, then

$$\int_j g(\mathbf{p}, \mathbf{q}) \, dq = 4h \log(1 + \sqrt{2}). \tag{16.2.6}$$

More general formulae for the integration of $g(\mathbf{p}, \mathbf{q})$ over rectangular and triangular elements, for any position of the point $\mathbf{p}$ (not necessarily in the plane of the element) are given by Birtles *et al.* (1973). Further related analytic integration formulae may be found in the work of Hess and Smith (1962).

As in two dimensions, the analytic results (16.2.4), in which the curvature of the true boundary surface ∂B is ignored, do not ensure the singular matrix approximation of the interior Neumann problem (see Chapter 13). Consequently, for such problems, we use instead the approximations

$$\int_j g'(\mathbf{q}_j, \mathbf{q})\,\mathrm{d}q = 2\pi - \sum_{\substack{i=1 \\ i \neq j}}^n g(\mathbf{q}_j, \mathbf{q}_i)' A_i \qquad (16.2.7)$$

and

$$\int_j g(\mathbf{q}_j, \mathbf{q})'\,\mathrm{d}q = 2\pi - \sum_{\substack{i=1 \\ i \neq j}}^n g(\mathbf{q}_j, \mathbf{q}_i)' A_i, \qquad (16.2.8)$$

corresponding to formulae (11.6.8) and (11.6.7) respectively.

16.3 Electrostatic capacitance

As a first application of an integral equation method in three dimensions, let us consider the problem of computing the electrostatic capacitance of a single charged conducting surface ∂B. Following the analysis of Section 2.3, we may formulate this problem either as a Dirichlet problem or as an interior Neumann problem. In the former case, the capacitance is given by

$$\kappa = \int_{\partial B} \lambda(\mathbf{q})\,\mathrm{d}q, \qquad (16.3.1)$$

where λ satisfies the equation

$$\int_{\partial B} g(\mathbf{p}, \mathbf{q})\, \lambda(\mathbf{q})\,\mathrm{d}q = 1, \qquad \mathbf{p} \in \partial B. \qquad (16.3.2)$$

In the latter case, the capacitance is given by

$$\kappa = \frac{1}{c} \int_{\partial B} \Lambda(\mathbf{q})\,\mathrm{d}q, \qquad (16.3.3)$$

where Λ is that solution of the homogeneous equation

$$\int_{\partial B} g'(\mathbf{p}, \mathbf{q})\,\Lambda(\mathbf{q})\,dq - 2\pi\Lambda(\mathbf{p}) = 0, \qquad \mathbf{p} \in \partial B, \tag{16.3.4}$$

which generates the constant interior potential c, i.e.

$$\int_{\partial B} g(\mathbf{p}, \mathbf{q})\,\Lambda(\mathbf{q})\,dq = c, \qquad \mathbf{p} \in B_i + \partial B. \tag{16.3.5}$$

In each case the capacitance is computed by discretising the relevant integral equation as described in Section 10.2,† solving the resulting system of simultaneous linear algebraic equations as described in Section 10.4, and thus approximating κ by a finite sum. In the Dirichlet formulation we denote this sum by

$$\tilde{\kappa}_D = \sum_{j=1}^{n} \lambda_j A_j. \tag{16.3.6}$$

In the Neumann case we take

$$\tilde{\kappa}_N = \sum_{j=1}^{n} \Lambda_j A_j / \tilde{c}, \tag{16.3.7}$$

where $\tilde{c}$ is an approximation to c obtained by averaging values computed from (16.3.5) at the nodal points.

Example 16.3.1

The problem: To determine the capacitance of a unit cube by (a) the Dirichlet formulation and (b) the Neumann formulation.

The method: Following Laskar (1971), we divide each face of the cube into m^2 equal square sub-areas where m is an odd positive integer. Then the centroid of each face is a nodal point and the total number of surface elements is $n = 6m^2$. The integral equations (16.3.2) and (16.3.4) are discretised using formula (16.2.3) in general together with either (16.2.6), with $h = 1/m$, or (16.2.7) at the nodal point $\mathbf{p} = \mathbf{q}_j$. From the symmetry of the problem, each

---

† Alternative treatments of equation (16.3.2), by variational methods, have been presented by Noble (1960) and McDonald *et al.* (1974).

equation reduces to a system of n^* simultaneous linear equations where $n^* = (m + 1)(m + 3)/8$. In case (b), the matrix of these equations is singular and a particular solution is obtained by setting $\Lambda_{n^*} = 1$ (in the corner element), omitting any one equation, and solving the remaining equations for $\Lambda_j; j = 1, 2, \ldots, n^* - 1$. In each case the algebraic equations are solved by the Gauss-Seidel method, described in Section 10.4, the iteration being terminated when components of successive solutions agree to within a preassigned increment ε. Finally, the capacitance of the cube is computed by means of either (a) (16.3.6) or (b) (16.3.7). In case (b), $\tilde{c}$ is obtained by discretising equation (16.3.5) in exactly the same manner as is equation (16.3.2) in case (a).

Results: Table 16.3.1 shows results obtained by Laskar (1974), taking $\varepsilon = 0.0001$ in the method described above.

TABLE 16.3.1

n	n^*	$\tilde{\kappa}_D$	$\tilde{\kappa}_N$
150	6	0·6538	0·6525
294	10	0·6568	0·6558
486	15	0·6582	0·6573
726	21	0·6590	0·6582
1014	28	0·6595	0·6588
1350	36	0·6598	0·6592
1734	45	0·6600	0·6595
2166	55	0·6602	0·6597
2646	66	0·6603	0·6599

Comments: It is interesting to note that the Dirichlet form of the solution presented here is essentially the same as the solution of Reitan and Higgins (1951) by the "method of subareas". More recently, Noble (1971) has obtained the value 0·6607 for the capacitance of the unit cube by a more accurate implementation of the Dirichlet formulation—dividing each face of the cube into rectangles of graded sizes and integrating the coefficients analytically.

Closely related to this problem is that of determining the capacitance of a thin square plate of unit area, for which Noble (1971) has obtained the value 0·367. Laskar (1974) obtained the result 0·362 by dividing the square into 289 equal elements yielding 45 independent equations, formed and solved as above. Also Birtles *et al.* (1973) chose this problem to illustrate the advantages of using a graded subdivision; they obtained an accuracy from 24 carefully chosen rectangles comparable with that from approximately 150 equal squares. Previously, results of comparable accuracy to Laskar's were obtained by Reitan and Higgins (1956), whose paper includes a clear account of earlier work in this field. Most of the references given here also cover rectangular plates with unequal sides, and Laskar (1974) considers triangular plates and circular discs too. An alternative treatment for the latter, with the rotational symmetry taken into account, will be described in Section 16.5.

16.4 Flow past an obstacle

As an example of an exterior Neumann problem in three dimensions, we consider once more the problem of potential flow past a fixed rigid obstacle. Thus, if ∂B is the surface of such an obstacle perturbing a steady flow defined by a velocity potential ψ, we seek the perturbation potential ϕ, in B_e, which satisfies the boundary condition

$$\phi'(\mathbf{p}) = -\psi'(\mathbf{p}), \qquad \mathbf{p} \in \partial B, \tag{16.4.1}$$

and which has the behaviour

$$\phi = O(r^{-2}) \quad \text{as} \quad r \to \infty. \tag{16.4.2}$$

Following the analysis of Section 2.4, we may represent this ϕ as a simple-layer Newtonian potential

$$\phi(\mathbf{p}) = \int_{\partial B} g(\mathbf{p}, \mathbf{q}) \, \sigma(\mathbf{q}) \, dq, \qquad \mathbf{p} \in B_e + \partial B, \tag{16.4.3}$$

in which case the source density σ is the unique solution of the integral equation

$$\int_{\partial B} g'(\mathbf{p}, \mathbf{q}) \, \sigma(\mathbf{q}) \, dq - 2\pi\sigma(\mathbf{p}) = -\psi'(\mathbf{p}), \qquad \mathbf{p} \in \partial B. \tag{16.4.4}$$

The two-dimensional equivalent of this formulation was applied to the problem of flow past an aerofoil in Section 13.6.

Alternatively, following the analysis of Section 3.3, we may apply Green's boundary formula (3.1.17) to ϕ. In this case, substituting for ϕ' from (16.4.1) into (3.3.5), we obtain the integral equation

$$\int_{\partial B} g(\mathbf{p}, \mathbf{q})' \, \phi(\mathbf{q}) \, dq - 2\pi\phi(\mathbf{p}) = -\int_{\partial B} g(\mathbf{p}, \mathbf{q}) \, \psi'(\mathbf{q}) \, dq, \qquad \mathbf{p} \in \partial B, \quad (16.4.5)$$

the solution of which yields ϕ on the surface ∂B directly.

Example 16.4.1

The problem: To determine the perturbation potential and the resultant velocity on the surface of a sphere in a uniform stream.

For a sphere placed in a stream flowing at constant speed U in the negative z-direction of a spherical coordinate system,

$$\psi = Uz = Ur \cos\theta \qquad (16.4.6)$$

in the analysis above, whence

$$\psi' = U \cos\theta \qquad (16.4.7)$$

when the origin of coordinates is taken at the centre of the sphere. Thus, from (16.4.1),

$$\phi' = -U \cos\theta \qquad (16.4.8)$$

and the analytic solution of this problem (Lamb, 1945) is given by

$$\phi = \tfrac{1}{2} \frac{Ua^3z}{r^3}, \qquad (16.4.9)$$

where a is the radius of the sphere. On the surface of the sphere the fluid velocity is, by rotational symmetry,

$$v_\theta = -\frac{1}{r}\frac{\partial}{\partial\theta}(\phi + \psi), \qquad (16.4.10)$$

which, from (16.4.6) and (16.4.9), becomes

$$v_\theta = \tfrac{3}{2} U \sin\theta. \qquad (16.4.11)$$

The method: Laskar (1971), whose results we quote below, has solved this problem by each of the two formulations above; first dividing the surface of the sphere into rings by planes orthogonal to the z-axis, and then approximating the spherical caps by six triangles and successive rings, approaching the equator, by increasing numbers of trapezoidal elements. The precise subdivision is a little complicated to describe and we shall not repeat the details here; we shall simply denote by n the total number of surface elements and by n^* the number of rings in one half of the sphere. Equations (16.4.4) and (16.4.5) are discretised by means of formulae (16.2.3), (16.2.4) and (16.2.5), and the resulting systems of n simultaneous linear equations are each reduced to n^* equations by taking account of symmetry. These equations are solved by the Gauss-Seidel method, described in Section 10.4, the iteration being terminated when components of successive solutions agree to within a preassigned increment ε. In the simple-layer potential formulation, the computed source density is used to generate an approximation $\tilde{\phi}_S$ to ϕ at the nodal points for comparison with the corresponding values $\tilde{\phi}_G$ obtained from Green's formula. For each formulation the tangential fluid velocity (16.4.10) is approximated at values of θ midway between the nodal values by simple finite difference formulae of the form

$$\frac{(\tilde{\phi} + \psi)_{j+1} - (\tilde{\phi} + \psi)_j}{a(\theta_{j+1} - \theta_j)}. \tag{16.4.12}$$

Results: For $U = 1$ and $a = 1$, results have been obtained as described

TABLE 16.4.1

θ (rad.)	ϕ	$\tilde{\phi}_S$	% error	$\tilde{\phi}_G$	% error
0·041	0·4996	0·5002	0·124	0·5020	0·478
0·096	0·4977	0·4992	0·293	0·5001	0·476
0·139	0·4952	0·4971	0·389	0·4976	0·478
0·179	0·4920	0·4941	0·433	0·4944	0·474
0·675	0·3905	0·3922	0·433	0·3921	0·423
0·707	0·3803	0·3819	0·434	0·3819	0·431
0·739	0·3697	0·3713	0·433	0·3714	0·441
0·771	0·3588	0·3603	0·435	0·3604	0·452
1·45	0·05794	0·05823	0·487	0·05829	0·594
1·49	0·04145	0·04166	0·487	0·04170	0·595
1·52	0·02489	0·02501	0·486	0·02504	0·598
1·55	0·00828	0·00832	0·486	0·00833	0·597

above with $n = 11112$, $n^* = 46$ and $\varepsilon = 0.0001$. Table 16.4.1 compares typical nodal values of $\tilde{\phi}_S$ and $\tilde{\phi}_G$ with the corresponding analytic values $\phi = \frac{1}{2}\cos\theta$ from (16.4.9). Figure 16.4.1 compares computed and analytic velocities on the surface of the sphere.

Comments: Although the error in $\tilde{\phi}_G$ is slightly larger than that in $\tilde{\phi}_S$ (Table 16.4.1), it is more uniform and consequently $\tilde{\phi}_G$ yields the more accurate surface velocity near the z-axis $\theta = 0$ (Fig. 16.4.1). It is perhaps surprising

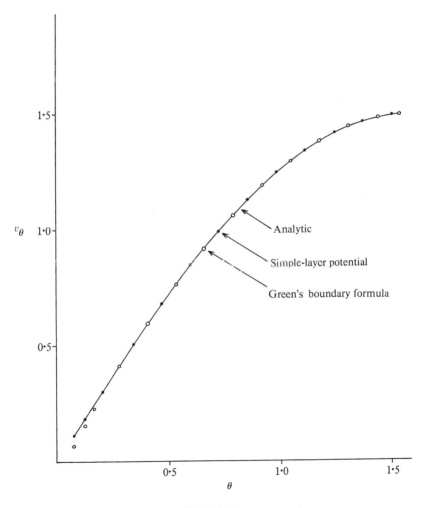

FIG. 16.4.1.

therefore that the Green's formulation has received so little attention in the past, in the field of fluid dynamics, whilst the simple-layer formulation has been extensively applied to potential flow problems. A comprehensive review of work in this area has been given by Hess (1975a).

One field in which the Green's formula has found practical application in three dimensions is in the application of electromagnetic theory to electrocardiography (Barnard *et al.*, 1967a, b). In connection with this work, Lynn and Timlake (1968a) have derived approximating systems of linear equations, for which they provide a rigorous error analysis; see also Ikebe *et al.* (1969) and, for multi-interface Neumann problems, Lynn and Timlake (1968b, 1970). We note also that Grodtkjær (1973) uses Green's formula to derive integral equations for the velocity components along the surface.

16.5 Problems with rotational symmetry

Problems with rotational symmetry can be solved more accurately than general three-dimensional problems by approximating their boundaries by ring-shaped elements rather than by planar polygons. In particular, we shall consider the Dirichlet problem for ϕ in a domain B bounded (internally or externally) by a surface of revolution ∂B, generated by rotating some plane contour C (not necessarily closed) about the z-axis. In this case, seeking ϕ in the form of a simple-layer potential (16.4.3) we must solve the integral equation

$$\int_{\partial B} g(\mathbf{p}, \mathbf{q})\sigma(\mathbf{q})\,\mathrm{d}q = \phi(\mathbf{p}), \qquad \mathbf{p} \in \partial B, \tag{16.5.1}$$

for σ, given ϕ on ∂B. Since ϕ is rotationally symmetric, it follows that σ also is rotationally symmetric.

Approximating the contour C by n chords, we approximate the surface ∂B by n rings $\Delta_1, \Delta_2, \ldots, \Delta_n$. A typical ring is the surface of a frustum of a cone generated by a chord AB as illustrated in Fig. 16.5.1. For a chord parallel to the z-axis, the corresponding ring is cylindrical, whilst for a chord orthogonal to this axis it becomes a circular disc or annulus. Then, approximating σ by a constant along each chord of C, and hence over each ring-shaped element Δ_j of ∂B, we approximate ϕ by

$$\tilde{\phi}(\mathbf{p}) = \sum_{j=1}^{n} \sigma_j \int_j g(\mathbf{p}, \mathbf{q})\,\mathrm{d}q; \qquad \mathbf{p} \in B + \partial B. \tag{16.5.2}$$

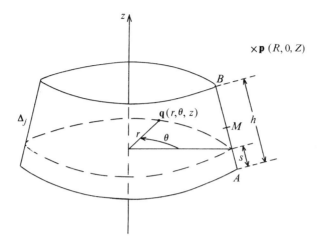

Fig. 16.5.1.

To evaluate the coefficient of σ_j in (16.5.2), we introduce cylindrical polar coordinates $\mathbf{q} = (r, \theta, z)$, $\mathbf{p} = (R, 0, Z)$ with no loss of generality. Then denoting by h the length of the generating chord AB in the plane $\theta = 0$ and by s the distance from A to $\mathbf{q}(\theta = 0)$, we find

$$I_j(\mathbf{p}) \equiv \int_j g(\mathbf{p}, \mathbf{q})\, dq = \int_j \frac{dq}{|\mathbf{p} - \mathbf{q}|}$$

$$= \int_0^h \left[\int_0^{2\pi} \frac{r\, d\theta}{\sqrt{\{(z - Z)^2 + (r + R)^2 - 4rR\cos^2(\theta/2)\}}} \right] ds. \qquad (16.5.3)$$

Defining

$$k^2 = \frac{4rR}{(z - Z)^2 + (r + R)^2}, \qquad (16.5.4)$$

we thus obtain

$$I_j(\mathbf{p}) = 2R^{-\frac{1}{2}} \int_0^h kr^{\frac{1}{2}} K(k)\, ds, \qquad (16.5.5)$$

where K is the complete elliptic integral of the first kind:

$$K(k) = \int_0^{\pi/2} \frac{d\alpha}{\sqrt{(1 - k^2 \sin^2 \alpha)}}, \qquad 0 \leqslant k < 1. \tag{16.5.6}$$

The latter integral may be approximated, with sufficient accuracy for our purposes, by

$$K(k) \doteqdot P(x) - Q(x) \log x; \qquad x = 1 - k^2, \tag{16.5.7}$$

where $P(x)$ and $Q(x)$ are second order polynomials (Hastings, 1955; Hart et al., 1968). Then, from (16.5.5), I_j may be evaluated by numerical quadrature. For example, subdividing AB into m equal intervals and applying the mid-ordinate rule (11.5.1), we obtain

$$I_j(\mathbf{p}) \doteqdot 2R^{-\frac{1}{2}} \frac{h}{m} \sum_{t=1}^{m} \left[kr^{\frac{1}{2}} K(k) \right]_{s=(t-\frac{1}{2})h/m}. \tag{16.5.8}$$

It may be observed from (16.5.7) that K has a logarithmic singularity as $x \to 0$ $(k \to 1)$, as a consequence of which the approximation (16.5.8) becomes inaccurate when the field point $\mathbf{p}$ lies close to the chord AB. In this case therefore, e.g. when $\mathbf{p}$ lies within distance h of the jth nodal point at the mid-point M of AB (Fig. 16.5.1), the integral I_j is obtained in two parts as follows.

A section $-\beta < \theta < \beta$ of the element Δ_j is treated as a rectangle of dimensions h by $2\beta r_M$, where r_M is the radial coordinate of M. In general, the integral of g over this rectangle may be obtained analytically (cf. Section 16.2) from the formula

$$a \log \left[\frac{b + \sqrt{(a^2 + b^2 + d^2)}}{\sqrt{(a^2 + d^2)}} \right] + b \log \left[\frac{a + \sqrt{(a^2 + b^2 + d^2)}}{\sqrt{(b^2 + d^2)}} \right]$$

$$- d \tan^{-1} \left[\frac{ab}{d\sqrt{(a^2 + b^2 + d^2)}} \right], \tag{16.5.9}$$

which gives the potential at perpendicular distance d above a corner of a rectangle of uniform unit source density of dimensions a by b. In the particular case when $\mathbf{p}$ coincides with M, the integral may be obtained from the analytic formula

$$4 \left\{ a \log \left[\frac{b + \sqrt{(a^2 + b^2)}}{a} \right] + b \log \left[\frac{a + \sqrt{(a^2 + b^2)}}{b} \right] \right\} \qquad (16.5.10)$$

for the potential at the centre of a rectangle of unit source density of dimensions $2a$ by $2b$. Formula (16.5.10), which follows immediately from (16.5.9), may also be derived from result (16.2.5).

The integral of g over the remainder of Δ_j is obtained by subtracting the integral

$$I_\beta(\mathbf{p}) = \int_{\pi/2 - \beta}^{\pi/2} \frac{d\alpha}{\sqrt{(1 - k^2 \sin^2 \alpha)}} \qquad (16.5.11)$$

from each elliptic integral in the quadrature formula (16.5.8). For small β

$$I_\beta(\mathbf{p}) \doteqdot \int_0^{\beta/2} \frac{d\alpha}{\sqrt{(1 - k^2 + k^2\alpha^2)}} = \frac{1}{k} \log \left[\frac{k\beta/2 + \sqrt{((k\beta/2)^2 + 1 - k^2)}}{\sqrt{(1 - k^2)}} \right]$$

$$(16.5.12)$$

which, for k near unity, may be further approximated by

$$I_\beta(\mathbf{p}) \doteqdot \frac{1}{k} \log \left[k\beta/2 + \sqrt{((k\beta/2)^2 + x)} \right] - 0 \cdot 5 \log x$$

$$- \frac{1}{k} \left[\frac{x}{4} + \frac{x^2}{16} \right] \log x; \qquad x = 1 - k^2. \qquad (16.5.13)$$

In the latter case, the term $-0 \cdot 5 \log x$ is cancelled out by a corresponding term in the expression (16.5.7) for K, and when $k = 1$ ($x = 0$) the only other contribution from (16.5.13) is $\log \beta$ from the first term. For k near zero, (16.5.12) yields a further approximation

$$I_\beta(\mathbf{p}) = \frac{\beta}{2}, \qquad (16.5.14)$$

which is applicable when the field point $\mathbf{p}$ lies close to the z-axis.

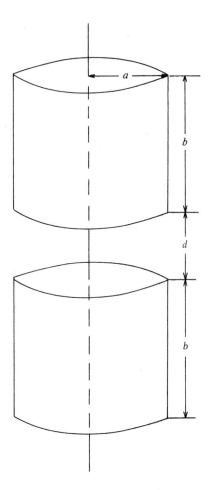

FIG. 16.5.2.

Example 16.5.1

The problem: To determine the electrostatic capacitance C between two solid circular cylinders, each of radius a and height b, separated by a distance d along a common axis, as illustrated in Fig. 16.5.2. If constant potentials ± 1 are imposed upon the two cylinders, then $C = Q/2$ where Q is the total absolute charge on either cylinder.

For small ratios b/a and d/a, this capacitance is given by Kirchhoff's approximation (Glazebrook, 1922)

$$C \doteq \frac{a}{3 \cdot 6}\left[\frac{a}{d} + \frac{1}{\pi}\left\{\log \frac{16\pi a(b + d)}{d^2} + \frac{b}{d}\log \frac{b + d}{b} - 1\right\}\right], \qquad (16.5.15)$$

the result being in picofarads if a, b and d are measured in centimetres.

The method: The formulation described above has been applied to this problem by means of a program due to Mautz and Harrington (1970). In this program, the approximation (16.5.7) is taken from Hart *et al.* (1968). The parameter m in the quadrature rule (16.5.8) is varied according to the distance of the field point $\mathbf{p}$ from the mid-point M of AB, and β is taken to be $\pi/40$. To obtain the results in practical units, the source density σ, which here represents electrostatic charge, is multiplied by $4\pi\varepsilon$ where $\varepsilon = 8\cdot85419 \times 10^{-12}$ (the permittivity of air in farad/metre). Correspondingly, for geometric data given in metres, the capacitance C is obtained in farads.

Results: Table 16.5.1 shows computed values $\tilde{C}$ of the capacitance C for a variety of geometries, comparing them with values C_K, obtained from Kirchhoff's formula (16.5.15), in certain cases. When $b \leqslant 5\cdot0$, the chord length $h = 1\cdot0$ for each surface element. For $b > 5\cdot0$, $h = 2\cdot0$ on the plane surfaces of each cylinder and varies from $1\cdot0$ to $4\cdot0$ on the curved surface. In particular, when $b = 50\cdot0$, this length is divided into ten intervals of length $4\cdot0$ with two intervals of lengths $3\cdot0$ and $2\cdot0$ at each end. In each case $m = 30, 25, 20, 10$ or 5 according as $|\mathbf{p}M| < h, h \leqslant |\mathbf{p}M| < 2h, 2h \leqslant |\mathbf{p}M| < 3h, 3h \leqslant |\mathbf{p}M| < 4h$ or $|\mathbf{p}M| \geqslant 4h$ respectively.

Comments: Results for $b = 0\cdot0$, the case of two thin discs, show how the calculated capacitance $\tilde{C}$ diverges from C_K as the discs' separation d increases. It is interesting to note that $\tilde{C}$ approaches the analytic limit $C = 0\cdot071$ for infinite separation (single disc), whereas C_K is already past this limit when $d = 40\cdot0$.

The numerical formulation employed here is essentially the same as that used by Higgins and Reitan (1951) in the calculation of the capacitance of a circular annulus by the method of subareas. Similar formulations for axisymmetric Neumann problems are reviewed in a paper by Hess (1975c) in connection with potential flow problems. An alternative formulation of the latter in terms of Green's boundary formula is demonstrated by Voinov *et al.* (1974).

TABLE 16.5.1

| Dimensions in metres | | | Capacitance in farads $\times 10^{-8}$ | |
a	b	d	$\tilde{C}$	C_K
20·0	0·0	4·0	0·361	0·358
		8·0	0·217	0·207
		12·0	0·167	0·153
		16·0	0·142	0·125
		20·0	0·127	0·107
		24·0	0·117	0·095
		28·0	0·110	0·085
		32·0	0·105	0·078
		36·0	0·100	0·072
20·0	0·0	40·0	0·097	0·067
10·0	1·0	1·0	0·340	0·336
10·0	1·0	2·0	0·194	0·187
10·0	5·0	2·0	0·215	
	10·0		0·228	
	20·0		0·251	
	30·0		0·269	
	40·0		0·284	
10·0	50·0	2·0	0·298	
10·0	50·0	0·2	1·566	
		0·4	0·868	
		0·6	0·634	
		0·8	0·516	
		1·0	0·445	
		2·0	0·298	
		5·0	0·201	
		10·0	0·163	
		15·0	0·148	
10·0	50·0	20·0	0·139	

APPENDICES

Appendix 1. Note on Liapunov and Kellogg Regular Surfaces

A Liapunov surface is a smooth surface possessing a tangent plane and normal, but not necessarily a curvature, at each point. This implies the existence of local coordinates at any point of the surface, z-axis along the normal, x- and y-axes in the tangent plane, such that a portion of the surface in this neighbourhood has the equation $z = z(x, y)$ where the partial derivatives z_x, z_y, but not necessarily z_{xx}, z_{yy}, z_{xy}, exist and are continuous.

If $\mathbf{n}_p, \mathbf{n}_q$ are the unit normal vectors at any two points $\mathbf{p}, \mathbf{q}$ of a Liapunov surface, then we require

$$\cos^{-1}(\mathbf{n}_p \cdot \mathbf{n}_q) \leqslant D|\mathbf{p} - \mathbf{q}|^v; \qquad 0 < v \leqslant 1, D > 0. \tag{1}$$

This condition is reminiscent of condition (1), Appendix 2, which characterises Hölder continuity. It has been proved by Shidfar (1977) that (1) holds for the surface defined by $z = z(x, y)$ if z_x, z_y are Hölder continuous over this surface. Symbolising a Liapunov surface by $S\{v\}$, and Hölder continuity similarly, Shidfar's result appears as

$$z_x \in H\{v_1\} \quad \text{and} \quad z_y \in H\{v_2\} \quad \text{imply} \quad z(x, y) \in S\{\min(v_1, v_2)\}. \tag{2}$$

Liapunov smoothness is stronger than $C^{(1)}$ smoothness. For instance the surface defined by

$$z = \int_0^x (\log t)^{-1} \, dt; \qquad 0 \leqslant x \leqslant \tfrac{1}{2}, \tag{3}$$

has $C^{(1)}$ smoothness. However, it does not satisfy condition (1) for any admissible choice of v, because z_x does not satisfy condition (1), Appendix 2, for any admissible choice of v.

In line with Hölder continuity, $v = 1$ defines the smoothest class of Liapunov surface. For instance

$$(x^2 + y^2)^{3/4} \in S\{\tfrac{1}{2}\} \quad \text{but} \quad \notin S\{1\}. \tag{4}$$

251

Any surface characterised by $C^{(2)}$ smoothness, e.g. the ellipsoid, belongs to $S\{1\}$, but the converse does not necessarily hold. Thus consider the surface ∂B formed by closing a finite circular cylinder by a hemisphere at each end. Clearly ∂B has a tangent plane and normal direction everywhere which vary continuously, and it may be readily proved that $\partial B \in S\{1\}$, but $\partial B \notin C^{(2)}$ since the curvature jumps on passing from the cylinder to the hemisphere.

Liapunov surfaces are less general than those considered by Kellogg, which could have corners or edges provided they were not too sharp. For instance, a cube is a Kellogg regular surface although a cone is not. A clear account of the distinction between Kellogg regular and Liapunov surfaces is given by Burton (1973).

Appendix 2. Hölder Continuity
{Lipschitz}

A function $f(x)$ satisfies a Hölder condition in the interval $a \leqslant x \leqslant b$, symbolised $f(x) \in H[a, b]$, if

$$|f(x_2) - f(x_1)| < D|x_2 - x_1|^\nu; \qquad 0 < \nu \leqslant 1, D > 0 \tag{1}$$

for any two distinct points $x_2, x_1 \in [a, b]$.

If $\nu - 1$, condition (1) becomes the Lipschitz condition, symbolised $f(x) \in L[a, b]$. Clearly the Lipschitz condition

$$|f(x_2) - f(x_1)| < D|x_2 - x_1| \tag{2}$$

implies (1) as $|x_2 - x_1| \to 0$, but not conversely, so that this condition defines the most restrictive class of Hölder continuous functions. Some writers refer to the Hölder condition as a Lipschitz condition, e.g. Smirnov (1964). It may also be remarked that $\nu > 1$ in (1) implies $f'(x) = 0$ for all $x \in [a, b]$.

If $f(x)$ is differentiable in $[a, b]$ then $f(x) \in L[a, b]$, but not conversely. For instance $|x| \in L[-\frac{1}{2}, \frac{1}{2}]$ but it is not differentiable at $x = 0$. This shows that differentiability is a stronger condition than Hölder continuity.

The function $f(x) = (\log x)^{-1}$, with $f(0) = 0$, is continuous in $[0, \frac{1}{2}]$ but it is not Hölder continuous, since we can always find x such that

$$|f(x) - f(0)| = |\log x|^{-1} > Dx^\nu; \qquad 0 < \nu \leqslant 1, D > 0 \tag{3}$$

irrespective of the choice of D, ν. Accordingly, Hölder continuity is a stronger condition than ordinary continuity. It may be noted that the function $f(x)$ is not differentiable at $x = 0$, though it is differentiable everywhere else in the interval, so that no contradiction arises with the previous paragraph.

Appendix 3. Uniqueness of Solutions for a Dielectric System

From (3.1.11) applied to ϕ_i and (3.1.19) applied to ϕ_e, we have

$$\int_{\partial B} \phi_i K_i \phi_i' \, dq = -K_i \int_{B_i} |\nabla \phi_i|^2 \, dQ, \tag{1}$$

$$\int_{\partial B} \phi_e K_e \phi_e' \, dq = -K_e \int_{B_e} |\nabla \phi_e|^2 \, dQ, \tag{2}$$

whence

$$\int_{\partial B} \phi(K_i \phi_i' + K_e \phi_e') \, dq = -K_i \int_{B_i} |\nabla \phi_i|^2 \, dQ - K_e \int_{B_e} |\nabla \phi_e|^2 \, dQ \tag{3}$$

on bearing in mind that $\phi_i = \phi_e = \phi$ on ∂B. In the absence of an external potential ψ in (2.4.16), the condition $K_i \phi_i' + K_e \phi_e' = 0$ holds on ∂B in which case

$$K_i \int_{B_i} |\nabla \phi_i|^2 \, dQ + K_e \int_{B_e} |\nabla \phi_e|^2 \, dQ = 0. \tag{4}$$

Since $K_i, K_e > 0$, a straightforward inference from (4) is that

$$\nabla \phi_i = 0 \quad \text{in} \quad B_i, \text{i.e. } \phi_i = C_i \text{ (a const.)} \quad \text{in} \quad B_i + \partial B,$$

$$\nabla \phi_e = 0 \quad \text{in} \quad B_e, \text{i.e. } \phi_e = C_e \text{ (a const.)} \quad \text{in} \quad B_e + \partial B.$$

Now $C_e = 0$ since $\phi_e = O(r^{-1})$ as $r \to \infty$, and therefore $C_i = 0$ since $\phi_i = \phi_e$ on ∂B. Accordingly $\phi = 0$ everywhere when $\psi = 0$. In this case equation (2.4.17) for σ in terms of ψ reduces to the homogeneous equation

$$(K_i - K_e) \int_{\partial B} g_i'(\mathbf{p}, \mathbf{q}) \sigma(\mathbf{q}) \, dq - 2\pi(K_i + K_e)\sigma(\mathbf{p}) = 0; \qquad \mathbf{p} \in \partial B, \tag{5}$$

which can have no solution other than $\sigma = 0$ by virtue of the relations

$$\int_{\partial B} g(\mathbf{p}, \mathbf{q})\sigma(\mathbf{q}) \, dq = \phi(\mathbf{p}), \phi(\mathbf{p}) = 0; \qquad \mathbf{p} \in \partial B.$$

An electrostatic conductor is defined by $K_i = \infty$ (formally equivalent to $K_e = 0$) in which case (4) yields

$$\phi_i = C \text{ (an arbitrary constant) in } B_i + \partial B. \tag{6}$$

Accordingly we may envisage $\phi_i = 1$ in $B_i + \partial B$ when $\psi = 0$, as has of course been assumed in Section 2.3. The corresponding equation for σ is

$$\int_{\partial B} g_i'(\mathbf{p}, \mathbf{q})\sigma(\mathbf{q}) \, dq - 2\pi\sigma(\mathbf{p}) = 0; \qquad \mathbf{p} \in \partial B, \tag{7}$$

and this must have a non-trivial solution λ which satisfies the relation

$$\int_{\partial B} g(\mathbf{p}, \mathbf{q})\lambda(\mathbf{q}) \, dq = 1; \qquad \mathbf{p} \in \partial B, \text{ in line with the theory of Section 2.3.}$$

An alternative formulation of the dielectric problem is via Green's boundary formula. Thus, putting $\phi = \phi_i$ in relation (3.1.12) and multiplying by K_i, and putting $\phi = \phi_e$ in relation (3.1.17) and multiplying by K_e, we find by superposition that

$$(K_i - K_e) \int_{\partial B} g(\mathbf{p}, \mathbf{q})_i' \phi(\mathbf{q}) \, dq - \int_{\partial B} g(\mathbf{p}, \mathbf{q})[K_i\phi_i'(\mathbf{q}) + K_e\phi_e'(\mathbf{q})] \, dq$$

$$= 2\pi(K_i + K_e)\phi(\mathbf{p}); \qquad \mathbf{p} \in \partial B, \tag{8}$$

where $\phi_i = \phi_e = \phi$ on ∂B. Since

$$K_i\phi_i' + K_e\phi_e' = (K_e - K_i)\psi_i' \tag{9}$$

from (2.4.16), the second integral in (8) may be written

$$(K_e - K_i) \int_{\partial B} g(\mathbf{p}, \mathbf{q})\psi_i'(\mathbf{q}) \, dq; \qquad \mathbf{p} \in \partial B, \tag{10}$$

and equation (8) therefore becomes

$$(K_i - K_e) \int_{\partial B} g(\mathbf{p}, \mathbf{q})'_i \phi(\mathbf{q}) \, dq - 2\pi(K_i + K_e)\phi(\mathbf{p})$$

$$= (K_e - K_i) \int_{\partial B} g(\mathbf{p}, \mathbf{q})\psi'_i(\mathbf{q}) \, dq; \qquad \mathbf{p} \in \partial B. \tag{11}$$

This is a Fredholm integral equation of the second kind for ϕ in terms of ψ_i on ∂B. By virtue of

$$\int_{\partial B} g(\mathbf{p}, \mathbf{q})'_i \psi(\mathbf{q}) \, dq - \int_{\partial B} g(\mathbf{p}, \mathbf{q})\psi'_i(\mathbf{q}) \, dq = 2\pi\psi(\mathbf{p}); \qquad \mathbf{p} \in \partial B, \tag{12}$$

the right-hand side of equation (11) transforms into

$$(K_e - K_i) \left\{ \int_{\partial B} g(\mathbf{p}, \mathbf{q})'_i \psi(\mathbf{q}) \, dq - 2\pi\psi(\mathbf{p}) \right\}; \qquad \mathbf{p} \in \partial B, \tag{13}$$

so enabling the equation to be written

$$(K_i - K_e) \int_{\partial B} g(\mathbf{p}, \mathbf{q})'_i [\phi(\mathbf{q}) + \psi(\mathbf{q})] \, dq - 2\pi(K_i + K_e)[\phi(\mathbf{p}) + \psi(\mathbf{p})]$$

$$= -4\pi K_e \psi(\mathbf{p}); \qquad \mathbf{p} \in \partial B. \tag{14}$$

Equation (14) has been put forward by Phillips (1934) using a different line of argument, as a Fredholm integral equation of the second kind for $\phi + \psi$ on ∂B in terms of ψ on ∂B. A numerical solution of this equation has been attempted (Edwards and Van Bladel, 1961).

Putting $\psi'_i = 0$ in equation (11) or $\psi = C$ (an arbitrary constant) in equation (14) yields the homogeneous equation

$$\int_{\partial B} g(\mathbf{p}, \mathbf{q})'_i \phi(\mathbf{q}) \, dq - \frac{2\pi(K_i + K_e)}{K_i - K_e} \phi(\mathbf{p}) = 0; \qquad \mathbf{p} \in \partial B. \tag{15}$$

This has the transpose form to equation (5). According to arguments put forward by Kellogg (1929), the inequality $(K_i + K_e)/(K_i - K_e) > 1$ precludes a non-trivial solution—in line with the theory of equation (5). When $K_i = \infty$ (or $K_e = 0$) equation (15) becomes the transpose to equation (7), with a non-trivial solution $\phi = 1$ to match the solution $\sigma = \lambda$ of equation (7).

Appendix 4. Poisson's Integral

Given a harmonic function ϕ within the sphere $r \leqslant a$, and a regular exterior harmonic function f defined by $f = -\phi$ on $r = a$, then f uniquely exists in $r \geqslant a$ and $f'_e + \phi'_i = \phi/a$ on $r = a$. This latter property may be proved without difficulty by utilising the expansion (3.6.1) for ϕ and the corresponding expansion for f. On introducing f into the continuation formula (3.4.2), we find

$$4\pi\phi(\mathbf{p}) = 2 \int_{\partial B} g(\mathbf{p}, \mathbf{q})'_i \phi(\mathbf{q}) \, dq - a^{-1} \int_{\partial B} g(\mathbf{p}, \mathbf{q}) \phi(\mathbf{q}) \, dq$$

$$= - \int_{\partial B} \{2g(\mathbf{p}, \mathbf{q})'_e + a^{-1} g(\mathbf{p}, \mathbf{q})\} \phi(\mathbf{q}) \, dq; \qquad \mathbf{p} \in B_i, \qquad (1)$$

for the sphere. Without loss of generality we may introduce the coordinates

$$\mathbf{p} = (0, 0, \rho), \quad \mathbf{q} = (r \sin\theta \cos\psi, \, r \sin\theta \sin\psi, \, r \cos\theta)_{r=a},$$

in which case

$$g(\mathbf{p}, \mathbf{q}) = |\mathbf{p} - \mathbf{q}|^{-1} = (\rho^2 + r^2 - 2\rho r \cos\theta)^{-1/2}_{r=a},$$

$$g(\mathbf{p}, \mathbf{q})'_e = \left[\frac{\partial}{\partial r} (\rho^2 + r^2 - 2\rho r \cos\theta)^{-1/2} \right]_{r=a}$$

$$= \frac{\rho \cos\theta - a}{(\rho^2 + a^2 - 2\rho a \cos\theta)^{3/2}},$$

$$2g(\mathbf{p}, \mathbf{q})'_e + a^{-1} g(\mathbf{p}, \mathbf{q}) = \frac{|\mathbf{p}|^2 - a^2}{a|\mathbf{p} - \mathbf{q}|^3}; \qquad |\mathbf{p}| = \rho. \qquad (2)$$

Substituting from (2) into (1) gives Poisson's integral

$$4\pi\phi(\mathbf{p}) = \frac{a^2 - |\mathbf{p}|^2}{a} \int_{\partial B} \frac{\phi(\mathbf{q})\, dq}{|\mathbf{p} - \mathbf{q}|^3}; \qquad \mathbf{p} \in B_i. \tag{3}$$

The double-layer integral in (1) jumps by an amount $-4\pi\phi(\mathbf{p})$ at $\mathbf{p} \in \partial B$, and therefore the right-hand side yields a null value at $\mathbf{p} \in \partial B$. At first sight the same holds for Poisson's integral, since $|\mathbf{p}| = a$ when $\mathbf{p} \in \partial B$. However a detailed limiting analysis shows that

$$\frac{a^2 - |\mathbf{p}|^2}{a} \int_{\partial B} \frac{\phi(\mathbf{q})\, dq}{|\mathbf{p} - \mathbf{q}|^3} \to 4\pi\phi(\mathbf{p}) \quad \text{as} \quad \mathbf{p} = (0, 0, \rho) \to (0, 0, a). \tag{4}$$

Briefly, we write $dq = a^2 \sin\theta\, d\theta d\psi$ and note that

$$\int_{\partial B_0} \frac{\phi(\mathbf{q})\, dq}{|\mathbf{p} - \mathbf{q}|^3} = 2\pi a^2 \phi(0, 0, a) \int_0^\varepsilon \frac{\sin\theta\, d\theta}{(\rho^2 + a^2 - 2\rho a \cos\theta)^{3/2}}$$

over a spherical cap ∂B_0 centred upon $(0, 0, a)$, sufficiently small to justify the approximation $\phi(\mathbf{q}) = \phi(0, 0, a)$ for any $\mathbf{q}$ within this cap. It follows without difficulty that

$$\frac{a^2 - \rho^2}{a} \int_{\partial B_0} \frac{\phi(\mathbf{q})\, dq}{|\mathbf{p} - \mathbf{q}|^3} \to 4\pi\phi(0, 0, a) \quad \text{as} \quad (0, 0, \rho) \to (0, 0, a), \tag{5}$$

since the integral contains a term $[a\rho(a - \rho)]^{-1}$. Also

$$\frac{a^2 - \rho^2}{a} \int_{\partial B - \partial B_0} \frac{\psi(\mathbf{q})\, dq}{|\mathbf{p} - \mathbf{q}|^3} \to 0 \quad \text{as} \quad (0, 0, \rho) \to (0, 0, a) \tag{6}$$

since the integral remains finite when $\mathbf{p} = (0, 0, a)$. The superposition of (5) and (6) gives (4).

In the case of a circle of radius a, we impose $f = O(1)$ at infinity, in which case the exterior formula (4.4.10) applies with

$$k \int_{\partial B} \lambda(\mathbf{p})\, dp = -2\pi \int_{\partial B} f(\mathbf{p})\lambda(\mathbf{p})\, dp, \qquad \text{i.e. } k = -a^{-1} \int_{\partial B} f(\mathbf{p})\, dp$$

K

since λ is now a constant. Accordingly, in place of (1),

$$
-2\pi\phi(\mathbf{p}) = 2 \int_{\partial B} g(\mathbf{p}, \mathbf{q})'_i \phi(\mathbf{q})\, dq + a^{-1} \int_{\partial B} \phi(\mathbf{q})\, dq
$$

$$
= -\int_{\partial B} \{2g(\mathbf{p}, \mathbf{q})'_e - a^{-1}\} \phi(\mathbf{q})\, dq, \tag{7}
$$

bearing in mind that now $f'_e + \phi'_i = 0$. Introducing the coordinates

$$
\mathbf{p} = (\rho, 0), \mathbf{q} = (r \cos \theta, r \sin \theta)_{r=a},
$$

we obtain

$$
g(\mathbf{p}, \mathbf{q}) = \tfrac{1}{2}\log (r^2 + \rho^2 - 2\rho r \cos \theta)_{r=a},
$$

$$
g(\mathbf{p}, \mathbf{q})'_e = \left[\frac{1}{2} \frac{\partial}{\partial r} \log (r^2 + \rho^2 - 2\rho r \cos \theta) \right]_{r=a} = \frac{a - \rho \cos \theta}{a^2 + \rho^2 - 2\rho a \cos \theta},
$$

$$
2g(\mathbf{p}, \mathbf{q})'_e - a^{-1} = \frac{a^2 - |\mathbf{p}|^2}{a|\mathbf{p} - \mathbf{q}|^2}; \qquad |\mathbf{p}| = \rho.
$$

Substituting into (7) gives the two-dimensional Poisson's formula

$$
2\pi\phi(\mathbf{p}) = \frac{a^2 - |\mathbf{p}|^2}{a} \int_{\partial B} \frac{\phi(\mathbf{q})\, dq}{|\mathbf{p} - \mathbf{q}|^2}; \qquad \mathbf{p} \in B_i. \tag{9}
$$

As before, it may be proved that

$$
\frac{a^2 - |\mathbf{p}|^2}{a} \int_{\partial B} \frac{\phi(\mathbf{q})\, dq}{|\mathbf{p} - \mathbf{q}|^2} \to 2\pi\phi(\mathbf{p}) \quad \text{as} \quad \mathbf{p} = (\rho, 0) \to (a, 0). \tag{10}
$$

There is no difficulty in demonstrating that formulae (3) and (9) hold for the respective exterior domains and that they give the right behaviour at infinity in each case.

Appendix 5. Hilbert's Integral Formula

Green's two-dimensional boundary formula (4.4.3) simplifies considerably for a circle. Thus, if

$$\mathbf{p} = (a \cos \theta, a \sin \theta), \qquad \mathbf{q} = (r \cos \alpha, r \sin \alpha)_{r=a}$$

are points on the circle $r = a$, then

$$|\mathbf{p} - \mathbf{q}| = \{r^2 + a^2 - 2ra \cos (\theta - \alpha)\}^{\frac{1}{2}}_{r=a} = 2a \sin \left| \frac{\theta - \alpha}{2} \right|,$$

$$g(\mathbf{p}, \mathbf{q}) = \log |\mathbf{p} - \mathbf{q}| = \log \left\{ 2a \sin \left| \frac{\theta - \alpha}{2} \right| \right\}$$

$$g(\mathbf{p}, \mathbf{q})'_i = \left(-\frac{\partial g}{\partial r} \right)_{r=a} = -\frac{1}{2a}.$$

Accordingly (4.4.3) becomes

$$-\frac{1}{2a} \int_{\partial B} \phi(\mathbf{q}) \, dq - \int_{\partial B} \phi'_i(\mathbf{q}) \log \left\{ 2a \sin \left| \frac{\theta - \alpha}{2} \right| \right\} dq = -\pi \phi(\mathbf{p}); \qquad \mathbf{p} \in \partial B,$$

$$(1)$$

where ϕ is a harmonic function in $B + \partial B$, i.e. in $r \leqslant a$. Utilising polar coordinates we have $\phi(\mathbf{p}) = \phi(r, \theta)$ for $\mathbf{p} \in B$, $\phi(\mathbf{p}) = \phi(a, \theta)$ for $\mathbf{p} \in \partial B$, $\phi(\mathbf{q}) = \phi(a, \alpha)$, $\phi'_i(\mathbf{q}) = (-\partial \phi / \partial r)_{r=a}$, $dq = a \, d\alpha$. Differentiating (1) with respect to θ, bearing in mind that the first integral is a constant, we find

$$\frac{1}{2\pi} \int_0^{2\pi} \left(r \frac{\partial \phi}{\partial r} \right)_{r=a} \cot \left(\frac{\theta - \alpha}{2} \right) d\alpha = -\left(\frac{\partial \phi}{\partial \theta} \right)_{r=a}. \qquad (2)$$

Now $r \, \partial \phi / \partial r$, $- \partial \phi / \partial \theta$ form a pair of conjugate harmonic functions in $r \leqslant a$. Conversely, any pair of conjugate harmonic functions (except constants) may be written in this form. Therefore (2) is essentially Hilbert's integral formula

$$\frac{1}{2\pi} \int_0^{2\pi} u(a, \alpha) \cot\left(\frac{\theta - \alpha}{2}\right) d\alpha = v(a, \theta), \tag{3}$$

connecting the boundary values of a pair of conjugate harmonic functions u, v which exist in $r \leqslant a$ (with the proviso that $v = 0$ corresponds to $u = 1$). The functions

$$\phi(r, \theta) = \left(\frac{r}{a}\right)^n \frac{\cos}{\sin} n\theta; \qquad n = 1, 2, \ldots \tag{4}$$

are harmonic in $r \leqslant a$, and

$$\phi(a, \theta) = \frac{\cos}{\sin} n\theta, \qquad \left(\frac{\partial \phi}{\partial r}\right)_{r=a} = \frac{n}{a} \frac{\cos}{\sin} n\theta. \tag{5}$$

Substituting from (5) into (1) yields the definite integrals

$$\int_0^{2\pi} \frac{\cos}{\sin} n\alpha \log\left\{2a \sin\left|\frac{\theta - \alpha}{2}\right|\right\} d\alpha = \frac{-\pi}{n} \frac{\cos}{\sin} n\theta; \qquad n = 1, 2, \ldots. \tag{6}$$

Clearly $\cos n\theta$, $\sin n\theta$ are independent non-trivial solutions of the homogeneous Fredholm integral equation

$$\int_0^{2\pi} \phi(\alpha) \log\left\{2a \sin\left|\frac{\theta - \alpha}{2}\right|\right\} d\alpha + \frac{\pi}{n} \phi(\theta) = 0; \qquad n = 1, 2, \ldots. \tag{7}$$

Alternatively expressed, they are independent eigenfunctions of the integral operator

$$\int_0^{2\pi} \ldots \log\left\{2a \sin\left|\frac{\theta - \alpha}{2}\right|\right\} d\alpha$$

appertaining to the eigenvalue $-\pi/n$ chosen from the set defined by $n = 1, 2, \ldots$.

These ideas may be readily extended to three-dimensional harmonic functions. Thus, if

$$\mathbf{p} = (a \sin \theta \cos \psi, a \sin \theta \sin \psi, a \cos \theta),$$

$$\mathbf{q} = (r \sin \alpha \cos \beta, r \sin \alpha \sin \beta, r \cos \alpha)_{r=a}$$

are points on the sphere $r = a$, then

$$|\mathbf{p} - \mathbf{q}| = (a^2 + r^2 - 2ar \cos \Delta)^{\frac{1}{2}}_{r=a} = \sqrt{2}a (1 - \cos \Delta)^{\frac{1}{2}},$$

$$g(\mathbf{p}, \mathbf{q}) = |\mathbf{p} - \mathbf{q}|^{-1} = (\sqrt{2}a)^{-1} (1 - \cos \Delta)^{-\frac{1}{2}}$$

$$g(\mathbf{p}, \mathbf{q})'_i = \left(-\frac{\partial g}{\partial r} \right)_{r=a} = (2\sqrt{2}a^2)^{-1} (1 - \cos \Delta)^{-\frac{1}{2}},$$

where $\Delta = \cos^{-1} \left(\dfrac{\mathbf{p} \cdot \mathbf{q}}{|\mathbf{p}||\mathbf{q}|} \right)$.

Accordingly, Green's three-dimensional boundary formula (3.1.12) becomes

$$\int_{\partial B} \frac{\phi(\mathbf{q}) \, dq}{2\sqrt{2}\, a^2 (1 - \cos \Delta)^{\frac{1}{2}}} - \int_{\partial B} \frac{\phi'(\mathbf{q}) \, dq}{\sqrt{2}a (1 - \cos \Delta)^{\frac{1}{2}}} = 2\pi\phi(\mathbf{p}); \qquad \mathbf{p} \in \partial B, \quad (8)$$

where ϕ is a harmonic function in $B + \partial B$, i.e. in $r \leqslant a$. Utilising spherical polar coordinates we have $\phi(\mathbf{p}) = \phi(r, \theta, \psi)$ for $\mathbf{p} \in B$, $\phi(\mathbf{p}) = \phi(a, \theta, \psi)$ for $\mathbf{p} \in \partial B$, $\phi(\mathbf{q}) = \phi(a, \alpha, \beta)$, $\phi'_i(\mathbf{q}) = (-\partial\phi/\partial r)_{r=a}$, $dq = a^2 \sin \alpha \, d\alpha \, d\beta$. Equation (8) is the analogue of (1). Nothing would seem to be gained by differentiating (8) with respect to θ or ψ and so arriving at the analogue of (3). The three-dimensional analogues of (4) are

$$\phi(r, \theta, \psi) = \left(\frac{r}{a} \right)^n S_n(\theta, \psi); \qquad n = 0, 1, 2, \ldots, \qquad (9)$$

where

$$S_n(\theta, \psi) = P_n^m (\cos \theta) \, \genfrac{}{}{0pt}{}{\cos}{\sin} \, m\psi; \qquad m = 0, 1, \ldots, n. \qquad (10)$$

Noting that

$$\phi(a, \theta, \psi) = S_n(\theta, \psi), \qquad \left(\frac{\partial \phi}{\partial r}\right)_{r=a} = \frac{n}{a} S_n(\theta, \psi), \qquad (11)$$

and substituting from (11) into (8), we obtain

$$\int_0^\pi \int_0^{2\pi} \frac{S_n(\alpha, \beta) \sin \alpha \, d\alpha \, d\beta}{\sqrt{2}(1 - \cos \Delta)^{\frac{1}{2}}} = \frac{4\pi}{2n + 1} S_n(\theta, \psi); \qquad n = 1, 2, \ldots \quad (12)$$

which are the analogues of (6). Clearly the angular harmonic functions (10) comprise $2n + 1$ independent non-trivial solutions of the homogeneous Fredholm integral equation

$$\int_0^\pi \int_0^{2\pi} \frac{\phi(\alpha, \beta) \sin \alpha \, d\alpha \, d\beta}{\sqrt{2}(1 - \cos \Delta)^{\frac{1}{2}}} - \frac{4\pi}{2n + 1} \phi(\theta, \psi) = 0. \qquad (13)$$

Alternatively expressed, they are independent eigenfunctions of the integral operator

$$\int_0^\pi \int_0^{2\pi} \cdots \frac{\sin \alpha \, d\alpha \, d\beta}{\sqrt{2}(1 - \cos \Delta)^{\frac{1}{2}}}$$

appertaining to the eigenvalue $4\pi/(2n + 1)$. We conclude by noting that the integral (12) may be evaluated directly as

$$\lim_{h \to 1} \int_0^\pi \int_0^{2\pi} \frac{S_n(\alpha, \beta) \sin \alpha \, d\alpha \, d\beta}{(1 - 2h \cos \Delta + h^2)^{\frac{1}{2}}},$$

utilising the expansions

$$(1 - 2h \cos \Delta + h^2)^{-\frac{1}{2}} = \sum_{n=0}^{\infty} h^n P_n(\cos \Delta); \qquad |h| < 1,$$

$$P_n(\cos \Delta) = P_n(\cos \theta) P_n(\cos \alpha) + 2 \sum_{m=1}^{n} \frac{(n - m)!}{(n + m)!}$$

$$P_n^m(\cos \theta) P_n^m(\cos \alpha) \cos m(\psi - \beta),$$

and the principle of dominated convergence (Sterling, 1977).

Appendix 6. Singular Elastostatic Displacement

Let $\mathscr{L}$ be a linear differential operator, acting at $\mathbf{q} \in B$, which transforms the displacement vector $g(\mathbf{q}_\eta, \mathbf{p}_\alpha)$; $\alpha = 1, 2, 3$ into the vector

$$\mathscr{L}g(\mathbf{q}_\eta, \mathbf{p}_\alpha); \qquad \alpha = 1, 2, 3. \tag{1}$$

The accompanying stress tensor is

$$\mathscr{L}G(\mathbf{q}_\eta, \mathbf{p}_{\alpha\beta}) = \mathscr{L}g(\mathbf{q}_\eta, \mathbf{p}_{\gamma/\delta})\, C^{\gamma\delta\alpha\beta}; \qquad \alpha, \beta = 1, 2, 3 \tag{2}$$

bearing in mind that the operations at $\mathbf{q}$ and $\mathbf{p}$ commute. This tensor satisfies the equations

$$\frac{\partial}{\partial p_\beta} \mathscr{L}G(\mathbf{q}_\eta, \mathbf{p}_{\alpha\beta}) = \mathscr{L}G(\mathbf{q}_\eta, \mathbf{p}_{\alpha\beta/\beta}) = -4\pi\, \mathscr{L}\delta(\mathbf{q}_\eta, \mathbf{p}_\alpha),$$

i.e.
$$\frac{\partial}{\partial p_\beta} \mathscr{L}G(\mathbf{q}_\eta, \mathbf{p}_{\alpha\beta}) = 0 \quad \text{if} \quad \mathbf{p} \neq \mathbf{q} \quad \text{or} \quad \alpha \neq \eta, \tag{3}$$

which are Cauchy's equilibrium equations everywhere except at the source point $\mathbf{q}$, showing that (1) is an elastostatic displacement vector everywhere except at the source point $\mathbf{q}$.

The traction vector associated with (2) at $\mathbf{p} \in \partial B$ is

$$\mathscr{L}g(\mathbf{q}_\eta, \mathbf{p}_\alpha)^* = \mathscr{L}G(\mathbf{q}_\eta, \mathbf{p}_{\alpha\beta})\, n_\beta; \qquad \alpha = 1, 2, 3. \tag{4}$$

This satisfies the integral relations

$$\int_{\partial B} \mathscr{L}g(\mathbf{q}_\eta, \mathbf{p}_\alpha)^*\, a_\alpha\, \mathrm{d}p = 4\alpha\, \mathscr{L}a_\eta, \tag{5}$$

$$\int_{\partial B} \mathscr{L}g(\mathbf{q}_\eta, \mathbf{p}_\alpha)^*(\mathbf{b} \wedge \mathbf{p})_\alpha\, \mathrm{d}p = 4\pi\, \mathscr{L}(\mathbf{b} \wedge \mathbf{q})_\eta, \tag{6}$$

where **a, b** are arbitrary constant vectors. These results follow from Section 5.4, assuming that $\mathscr{L}$ commutes with the integral operator. Expressed in dyadic symbolism, relations (5) and (6) appear as

$$\int_{\partial B} \mathscr{L}\, \mathbf{g}(\mathbf{q}, \mathbf{p})^* \cdot \mathbf{a}\; dp = 4\pi\mathscr{L}\mathbf{a}, \tag{7}$$

$$\int_{\partial B} \mathscr{L}\mathbf{g}(\mathbf{q}, \mathbf{p})^* \cdot (\mathbf{b} \wedge \mathbf{p})\; dp = 4\pi\mathscr{L}(\mathbf{b} \wedge \mathbf{q}). \tag{8}$$

The cases

$$\mathscr{L}\mathbf{g}(\mathbf{q}, \mathbf{p}) = \mathbf{q} \cdot \nabla\mathbf{g}(\mathbf{q}, \mathbf{p}), \qquad \mathscr{L}\mathbf{g}(\mathbf{q}, \mathbf{p}) = \mathbf{g}^*(\mathbf{q}, \mathbf{p})$$

are relevant to the asymptotic expansions of Section 6.1.

Appendix 7. Strain-Energy Density

The strain-energy density is defined to be

$$\tfrac{1}{2}\Phi_{\alpha\beta}\,\eta_{\alpha\beta}; \qquad \alpha, \beta = 1, 2, 3 \tag{1}$$

where $\Phi_{\alpha\beta}\,(=\Phi_{\beta\alpha})$ is the stress component associated with the strain component $\eta_{\alpha\beta}(=\eta_{\beta\alpha})$. Substituting the stress–strain relations

$$\Phi_{\alpha\beta} = C^{\alpha\beta\gamma\delta}\eta_{\gamma\delta}; \qquad \gamma, \delta = 1, 2, 3$$

into (1), we obtain a quadratic form

$$\tfrac{1}{2}C^{\alpha\beta\gamma\delta}\eta_{\gamma\delta}\eta_{\alpha\beta} \tag{2}$$

in the strain components. A convenient symbolism for studying (2) has been introduced by Voigt (1928), viz.

$$\eta_{11} = \eta_1, \qquad \eta_{22} = \eta_2, \qquad \eta_{33} = \eta_3$$

$$\eta_{23} + \eta_{32} = \eta_4, \qquad \eta_{31} + \eta_{13} = \eta_5, \qquad \eta_{12} + \eta_{21} = \eta_6$$

and correspondingly

$$C^{\alpha\beta\gamma\delta} = C^{(\alpha\beta)(\gamma\delta)} = C^{(\gamma\delta)(\alpha\beta)}$$

where

$$(\alpha\beta), (\gamma\delta) = 1, 2, \ldots, 6.$$

In terms of this symbolism, (2) appears (omitting the factor $\tfrac{1}{2}$) as

$$C^{(1)(1)}\eta_1\eta_1 + \ldots + 2C^{(1)(2)}\eta_1\eta_2 + \ldots + 2C^{(1)(4)}\eta_1\eta_4 + \ldots + 2C^{(5)(6)}\eta_5\eta_6,$$

$$\tag{3}$$

which is a quadratic form in the variables $\eta_1, \eta_2, \ldots, \eta_6$ characterised by a symmetric matrix of coefficients. The necessary and sufficient conditions (Tropper, 1969) for (3) to be positive–definite, as required for a stable elastic material, are

$$C^{(1)(1)} > 0, \begin{vmatrix} C^{(1)(1)} & C^{(1)(2)} \\ C^{(2)(1)} & C^{(2)(2)} \end{vmatrix} > 0, \ldots \begin{vmatrix} C^{(1)(1)} & \ldots & C^{(1)(6)} \\ \vdots & & \vdots \\ C^{(6)(1)} & \ldots & C^{(6)(6)} \end{vmatrix} > 0. \tag{4}$$

If these conditions are satisfied, then (3) can be transformed into the form

$$\xi_1^2 + \xi_2^2 + \ldots + \xi_6^2 \tag{5}$$

where $\xi_1, \xi_2, \ldots, \xi_6$ are linearly related to $\eta_1, \eta_2, \ldots, \eta_6$. Now (5) cannot be zero unless $\xi_1 = \xi_2 = \ldots = \xi_6 = 0$, which implies $\eta_1 = \eta_2 = \ldots = \eta_6 = 0$. Therefore, if (1) has the value zero, we may conclude that $\eta_{\alpha\beta} = 0; \alpha, \beta = 1, 2, \ldots, 6$.

Appendix 8. On a First-Order Differential Equation Satisfied by Harmonic Functions

Given a harmonic function f in B, we seek a second harmonic function ϕ in B which satisfies the differential equation

$$k\phi + x\frac{\partial\phi}{\partial x} + y\frac{\partial\phi}{\partial y} + z\frac{\partial\phi}{\partial z} = f \tag{1}$$

in B where k is a constant. Following Bergman and Schiffer (1953), we transform (1) into

$$k\phi + r\frac{\partial\phi}{\partial r} = f \tag{2}$$

where r is the radial distance of x, y, z from an origin within B. Equation (2) has the family of solutions

$$\phi = Cr^{-k} + r^{-k}\int^r f\rho^{k-1}\,d\rho, \tag{3}$$

where C is independent of r. It may be verified that

$$\nabla^2(r^{-k}\int^r f\rho^{k-1}\,d\rho) = 0, \tag{4}$$

which shows that the choice $C = 0$ in (3) defines a unique harmonic solution of (1). Alternatively, we may operate upon both sides of (1) with ∇^2 to obtain

$$(k+2)\nabla^2\phi + r\frac{\partial}{\partial r}(\nabla^2\phi) = 0, \tag{5}$$

which has the family of solutions

$$\nabla^2\phi = Dr^{-(k+2)} \tag{6}$$

269

where D is an arbitrary constant: the choice $D = 0$ in (6) implies the existence of harmonic solutions of equation (1).

A useful test harmonic function in B is

$$f = r^n P_n^m(\cos \theta) \, e^{im\psi}; \qquad n = 0, 1, 2, \ldots, \quad m = -n, \ldots, n \qquad (7)$$

in which case

$$\phi = r^{-k} \int^r \rho^{k+n-1} P_n^m(\cos \theta) \, e^{im\psi} \, d\rho = \frac{r^n}{k+n} P_n^m(\cos \theta) \, e^{im\psi}. \qquad (8)$$

This shows that ϕ may not exist if $k = 0, -1, -2, \ldots$. For instance equation (1) has no solution when $f = 1$, defined by $n = 0$ in (7), if $k = 0$. Equation (8.1.19) of the text corresponds with $k = -1/\kappa$, showing that a breakdown may occur if $1 - \kappa n = 0$. However a solution of equation (8.3.10) always exists since this corresponds with $k = 3/2$. The exterior analogue of (7) is

$$f = r^{n-1} P_n^m(\cos \theta) \, e^{im\psi}; \qquad n = 0, -1, -2, \ldots, \quad m = -n, \ldots, n, \qquad (9)$$

yielding the exterior breakdown possibilities $k = 1, 2, \ldots$. In two dimensions the appropriate test functions are

$$f = r^n \frac{\cos}{\sin} n\theta; \qquad n = 0, \pm 1, \pm 2, \ldots, \qquad (10)$$

yielding the breakdown possibilities $k = 0, \mp 1, \mp 2, \ldots$ of which $k = 1$ covers equation (8.4.5) of the text. Breakdown possibilities have been further explored by Shidfar (1977).

Appendix 9. Rayleigh–Green Identity

Given two functions ϕ, ψ which are continuous in $B_i + \partial B$ and differentiable to the fourth order in B_i, we introduce the volume integral

$$\int_{B_i} (\phi \nabla^4 \psi - \psi \nabla^4 \phi) \, dQ \tag{1}$$

which may be written

$$\int_{B_i} \{\phi \nabla^2(\nabla^2 \psi) - \nabla^2 \psi \nabla^2 \phi\} \, dQ - \int_{B_i} \{\psi \nabla^2(\nabla^2 \phi) - \nabla^2 \phi \nabla^2 \psi\} \, dQ. \tag{2}$$

Applying the reciprocal relation (3.1.4) to each of the integrals in (2) gives:–

'left': $\int_{B_i} \{\phi \nabla^2(\nabla^2 \psi) - \nabla^2 \psi \nabla^2 \phi\} \, dQ = - \int_{\partial B} \{\phi(\nabla^2 \psi)' - (\nabla^2 \psi) \phi'\} \, dq,$ (3)

'right': $\int_{B_i} \{\psi \nabla^2(\nabla^2 \phi) - \nabla^2 \phi \nabla^2 \psi\} \, dQ - = - \int_{\partial B} \{\psi(\nabla^2 \phi)' - (\nabla^2 \phi) \psi'\} \, dq,$ (4)

whence

$$\int_{B_i} (\phi \nabla^4 \psi - \psi \nabla^4 \phi) \, dQ = - \int_{\partial B} \{\phi(\nabla^2 \psi)' - (\nabla^2 \psi) \phi'\} \, dq$$

$$+ \int_{\partial B} \{\psi(\nabla^2 \phi)' - (\nabla^2 \phi) \psi'\} \, dq. \tag{5}$$

This is the Rayleigh–Green identity.
We now choose ϕ to be biharmonic in B_i, so that

$$\nabla^4 \phi(\mathbf{q}) = 0; \qquad \mathbf{q} \in B_i. \tag{6}$$

Also, we choose

$$\psi(\mathbf{q}) = |\mathbf{q} - \mathbf{p}| \equiv R; \qquad \mathbf{q} \in B_i + \partial B, \qquad \mathbf{p} \in B_i, \tag{7}$$

so that

$$\nabla^2 \psi(\mathbf{q}) = 2R^{-1}, \qquad \nabla^4 \psi(\mathbf{q}) - - 8\pi\delta(\mathbf{q} - \mathbf{p}). \tag{8}$$

Substituting (6), (7), (8) formally into (5) yields

$$-2 \int_{\partial B} \left\{ \phi \left(\frac{1}{R}\right)' - \left(\frac{1}{R}\right) \phi' \right\} dq + \int_{\partial B} \{ R(\nabla^2\phi)' - (\nabla^2\phi) R' \} \, dq$$

$$= -8\pi \int_{B_i} \phi(\mathbf{q}) \, \delta(\mathbf{q} - \mathbf{p}) \, dQ = -8\pi\phi(\mathbf{p}); \qquad \mathbf{p} \in B_i. \tag{9}$$

This is the biharmonic counterpart of Green's formula (3.1.3). A straightforward extension of the analysis yields

$$-2 \int_{\partial B} \left\{ \phi \left(\frac{1}{R}\right)' - \left(\frac{1}{R}\right) \phi' \right\} dq + \int_{\partial B} \{ R(\nabla^2\phi)' - (\nabla^2\phi) R' \} \, dq$$

$$= -4\pi\phi(\mathbf{p}); \qquad \mathbf{p} \in \partial B, \tag{10}$$

$$= 0; \qquad \mathbf{p} \in B_e. \tag{11}$$

It may be noted that relation (10) provides a functional constraint between the four boundary quantities ϕ, ϕ', $\nabla^2\phi$, $(\nabla^2\phi)'$. Also $\nabla^2\phi$, $(\nabla^2\phi)'$ satisfy the boundary formula (3.1.12) since $\nabla^2\phi$ is a harmonic function in B_i. In biharmonic boundary-value problems, either ϕ and ϕ' or ϕ and $\nabla^2\phi$ are prescribed on ∂B, so that (10) and (3.1.12) constitute two functional equations for the remaining two boundary quantities. By solving for these, we may determine all four boundary quantities and so generate ϕ in B_i from (9). There is no difficulty in adapting the theory to two dimensions, and this formulation has been applied to certain problems of thin elastic plates by Segedin and Brickell (1968) and by Brickell (1970).

The Rayleigh–Green identity has been invoked by Maiti and Chakrabarty (1974) in connection with the Chakrabarty representation.

References

Allen, D. N. de G. and Robins, B. (1962). The application of relaxation methods to satisfy normal-gradient boundary conditions associated with three-dimensional partial differential equations. *Quart. J. Mech. Appl. Math.* **15** (1) 43–51.

Almansi, E. (1897). Sull' integrazione dell' equazione differentiale $\Delta^{2n} = 0$. *Annali di matematica pura et applicata*, series III, **2**, 1–51.

Banerjee, P. K. (1976). Integral equation methods for analysis of piece-wise non-homogeneous three-dimensional elastic solids of arbitrary shape. *Int. J. Mech. Sci.* **18**, 293–303.

Barnard, A. C. L., Duck, I. M. and Lynn, M. S. (1967a). The application of electromagnetic theory to electrocardiography. I—Derivation of the integral equations. *Biophys. J.* **7**, 430–462.

Barnard, A. C. L., Duck, I. M., Lynn, M. S. and Timlake, W. P. (1967b). The application of electromagnetic theory to electrocardiography. II—Numerical solution of the integral equations. *Biophys. J.* **7**, 463–491.

Barone, M. R. and Robinson, A. R. (1972). Determination of elastic stresses at notches and corners by integral equations. *Int. J. Solids Structures*, **8**, 1319–1338.

Bergman, S. and Schiffer, M. (1953). "Kernel Functions and Elliptic Differential Equations in Mathematical Physics." Academic Press, New York and London.

Bernal, M. J. M. and Whiteman, J. R. (1970). Numerical treatment of biharmonic problems with re-entrant boundaries. *Computer J.* **13** (1) 87–91.

Besuner, P. M. and Snow, D. W. (1975). Application of the two-dimensional integral equation method to engineering problems. *In* "Boundary-Integral Equation Method: Computational Applications in Applied Mechanics," (T. A. Cruse and F. J. Rizzo, Eds.). ASME, New York.

Bhargava, R. D. (1969). "Some Problems of Two-Dimensional Elasto-Plastic Stress Fields." Ph.D. Thesis, University of London.

Bhattacharyya, P. (1975). "Biharmonic Potential Theory and its Applications." Ph.D. Thesis, The City University, London.

Birkhoff, G. (1971). "The Numerical Solution of Elliptic Equations." SIAM, Philadelphia.

Birtles, A. B., Mayo, B. J. and Bennett, A. W. (1973). Computer techniques for solving 3-dimensional electron-optics and capacitance problems. *Proc. IEE*, **120** (2), 213–220.

Bobrik, A. I. and Mikhailov, V. N. (1974). The solution of certain problems for Poisson's equation with boundary conditions of the fourth kind. *USSR Comp. Maths. Math. Phys.* **14** (1), 127–134.

Brickell, D. G. A. (1970). Integral equation method for influence surfaces of a corner plate. *Proc. Instn. Civ. Engrs.* **46**, 185–194.

Brown, I. C. and Jaswon, E. (1971). "The Clamped Elliptic Plate under a Concentrated Transverse Load." Research Memorandum No. 6, Department of Mathematics, The City University, London.

Burton, A. J. (1973). "The Solution of Helmholtz' Equation in Exterior Domains Using Integral Equations." NPL Report NAC 30. *Kellog regular & Liapunov surfaces*

Butterfield, R. (1971). The application of integral equation methods to continuum problems in soil mechanics. *In* "Stress-Strain Behaviour of Soils," Proc. Roscoe Mem. Symp. (R. H. G. Parry, Ed.). Foulis, Henley-on-Thames.

273

Chakrabarty, S. K. (1971). "Numerical Solutions of some Integral Equations Related to Biharmonic Problems." Ph.D. Thesis, Indian Institute of Technology, Kharagpur.

Christiansen, S. (1971). Numerical solution of an integral equation with a logarithmic kernel. *BIT*, **11** (3), 276–287.

Christiansen, S. (1975). Integral equations without a unique solution can be made useful for solving some plane harmonic problems. *J. Inst. Maths. Applics.* **16**, 143–159.

Christiansen, S. (1976). On Kupradze's functional equations for plane harmonic problems. Function Theoretic Methods in Differential Equations, edited by Gilbert, R. P. and Weinacht, R. J. Pitman, London.

Christiansen, S. and Rasmussen, H. (1976). Numerical solutions for two-dimensional annular electromechanical machining problems. *J. Inst. Maths. Applics.* **18**, 295–307.

Courant, R. and Hilbert, D. (1953). "Methods of Mathematical Physics," Vol. I. Interscience, New York.

Cruse, T. A. (1969). Numerical solutions in three dimensional elastostatics. *Int. J. Solids Structures*, **5**, 1259–1274.

Cruse, T. A. (1972a). Some classical elastic sphere problems solved numerically by integral equations. *J. Appl. Mech., Trans ASME*, **39** (1), 272–274.

Cruse, T. A. (1972b). Application of the boundary-integral equation solution method in solid mechanics. Paper presented at International Conference on Variational Methods in Engineering, Dept. Civil Engng., Southampton University, Sept. 25–29, 1972.

Cruse, T. A. (1973). Application of the boundary-integral equation method to three dimensional stress analysis. *Computers and Structures*, **3**, 509–527.

Cruse, T. A. (1974). An improved boundary-integral equation method for three dimensional elastic stress analysis. *Computers and Structures*, **4**, 741–754.

Cruse, T. A. (1975). Boundary-integral equation fracture mechanics analysis. In "Boundary-Integral Equation Method: Computational Applications in Applied Mechanics," (T. A. Cruse and F. J. Rizzo, Eds.). ASME, New York.

Cruse, T. A., Osias, J. R. and Wilson, R. B. (1976). "Boundary–Integral Equation Method for Elastic Fracture Mechanics Analysis." Report No. AFOSR-TR-76-0878, Pratt and Whitney Aircraft Group, East Hartford.

Cruse, T. A. and Rizzo, F. J. (Eds.) (1975). "Boundary-Integral Equation Method: Computational Applications in Applied Mechanics." ASME, New York.

de Wolf, S. and de Mey, G. (1976). Numerical solution of integral equations for potential problems by a variational principle. *Inform. Proc. Lett.* **4** (5), 136–139.

Dirac, P. A. M. (1935). "The Principles of Quantum Mechanics." Second edition. Clarendon, Oxford.

Duff, G. F. D. (1956). "Partial Differential Equations." University of Toronto Press.

Duffin, R. J. (1974). Some problems of mathematics and physics. *Bull. Amer. Math. Soc.* **80**, 1053–1070.

Edwards, T. W. and Van Bladel, J. (1961). Electrostatic dipole moment of a dielectric cube. *Appl. Sci. Res. (B)*, **9**, 151–155.

Eshelby, J. D. (1957). The determination of the elastic field of an ellipsoidal inclusion, and related problems. *Proc. Roy. Soc. (A)*, **241**, 376–396.

Eubanks, R. A. and Sternberg, E. (1956). On the completeness of the Boussinesq–Papkovich stress functions. *J. Rat. Mech. Anal.* **5** (5), 735–746.

Evans, G. C. (1927). "The Logarithmic Potential, Discontinuous Dirichlet and Neumann Problems." Vol. VI, Colloq. Pub. Amer. Math. Soc., New York.

Fox, L. (1944). Solution by relaxation methods of plane potential problems with mixed boundary conditions. *Quart. Appl. Math.* **2**, 251–257.

Fox, L. and Goodwin, E. T. (1953). The numerical solution of non-singular linear integral equations. *Phil. Trans. Roy. Soc. (A)*, **245**, 501–534.

Fredholm, I. (1903). Sur une classe d'équations fonctionelles. *Acta Math.* **27**, 365–390.

Furuhashi, R. (1973). On some basic properties of generalised solution in elastostatics. *Bull. JSME*, **16**, 1239–1246.

Gaier, D. (1964). "Konstruktive Methoden der konformen Abbildung." Springer, Berlin.

Gaier, D. (1976). Integralgleichungen erster Art und konforme Abbildung. *Math. Z.* **147**, 113–129.

Glazebrook, Sir R. (Ed.) (1922). "A Dictionary of Applied Physics," Vol. II: "Electricity." Macmillan, London.

Goldenberg, H. (1969a). External thermal resistance of two buried cables. Restricted application of superposition. *Proc. IEE*, **116** (5), 822–826.

Goldenberg, H. (1969b). External thermal resistance of three buried cables in trefoil-touching formation. *Proc. IEE*, **116** (11), 1885–1890.

Goldenberg, H. (1971). "External Thermal Resistance of Three Buried Cables in Close Flat Formation." ERA Report No. 71–8.

Golecki, J. J. (1974). On integration of differential equations in elastostatics through determination of the mean stress. *Aplikace Matematiky*, **19**, 293–306.

Grodtkjær, E. (1973). A direct integral equation method for the potential flow about arbitrary bodies. *Int. J. Num. Methods Engng.* **6**, 253–264.

Günter, N. M. (1967). "Potential Theory and its Applications to Basic Problems of Mathematical Physics." Ungar, New York.

Hadamard, J. (1908). Mémoire sur le problème d'analyse relatif à l'équilibre des plaques élastiques encastrées. *Mém. prés. div. sav., Acad. Sci. Inst. France*, **33** (4), 1–126.

Hamel, G. (1949). "Integralgleichungen." Springer-Verlag, Berlin.

Hansen, E. B. (1976). Numerical solution of integro-differential and singular integral equations for plate bending problems. *J. Elasticity*, **6** (1), 39–56.

Harrington, R. F. (1968). "Field Computation by Moment Methods." Macmillan, New York.

Harrington, R. F., Pontoppidan, K., Abrahamsen, P. and Albertsen, N. C. (1969). Computation of Laplacian potentials by an equivalent-source method. *Proc. IEE*, **116** (10), 1715–1720.

Hart, J. F., Cheney, E. W., Lawson, C. L., Maehly, H. J., Mesztenyi, C. K., Rice, J. R., Thatcher, H. G., Jr. and Witzgall, C. (1968). "Computer Approximations." Wiley, New York.

Hastings, C., Jr. (1955). "Approximations for Digital Computers." Princeton University Press.

Hayes, J. K. (1970). "Four Computer Programs Using Green's Third Formula to Numerically Solve Laplace's Equation in Inhomogeneous Media." Los Alamos Scientific Laboratory Report LA-4423.

Hayes, J. and Kellner, R. (1972). The eigenvalue problem for a pair of coupled integral equations arising in the numerical solution of Laplace's equation. *SIAM J. Appl. Math.* **22** (3), 503–513.

Hayes, J. K., Kahaner, D. K. and Kellner, R. G. (1972). An improved method for numerical conformal mapping. *Maths. Comp.* **26**, 118, 327–334.

Hayes, J. K., Kahaner, D. K. and Kellner, R. G. (1975). A numerical comparison of integral equations of the first and second kind for conformal mapping. *Maths. Comp.* **29** (130) 512–521.

Heap, B. R. (1972)."Algorithms for the Production of Contour Maps over an Irregular Triangular Mesh." NPL Report NAC 10.

Hess, J. L. (1973). Higher order numerical solution of the integral equation for the two-dimensional Neumann problem. *Computer Methods Appl. Mech. Engng.* **2**, 1–15.

Hess, J. L. (1974). The problem of three-dimensional lifting potential flow and its solution by means of surface singularity distribution. *Computer Methods Appl. Mech. Engng.* **4**, 283–319.

Hess, J. L. (1975a). Review of integral-equation techniques for solving potential-flow problems with emphasis on the surface-source method. *Computer Methods Appl. Mech. Engng.* **5**, 145–196.

Hess, J. L. (1975b). The use of higher-order surface singularity distributions to obtain improved potential flow solutions for two-dimensional lifting aerofoils. *Computer Methods Appl. Mech. Engng.* **5**, 11–35.

Hess, J. L. (1975c). Improved solution for potential flow about arbitrary axisymmetric bodies by the use of a higher-order surface source method. *Computer Methods Appl. Mech. Engng.* **5**, 297–308.

Hess, J. L. and Smith, A. M. O. (1962). "Calculation of Non-Lifting Potential Flow About Arbitrary Three-Dimensional Bodies." Report No. E. S. 40622, Douglas Aircraft Co., Long Beach.

Hess, J. L. and Smith, A. M. O. (1967). Calculation of potential flow about arbitrary bodies. *In* "Progress in Aeronautical Sciences," Volume 8, (D. Kuchemann, Ed.). Pergamon, London.

Higgins, T. J. and Reitan, D. K. (1951). Calculation of the capacitance of a circular annulus by the method of subareas. *AIEE Trans.* **70** (1), 926–933.

Householder, A. S. (1964). "The Theory of Matrices in Numerical Analysis." Blaisdell, New York.

Ikebe, Y., Lynn, M. S. and Timlake, W. P. (1969). The numerical solution of the integral equation formulation of the single interface Neumann problem. *SIAM J. Numer. Anal.* **6**, 334–346.

Jaswon, E. (1973). "The Integral Equation Approach to Thin Plate Problems." Ph.D. Thesis, Technion—Israel Institute of Technology, Haifa.

Jaswon, M. A. (1963). Integral equation methods in potential theory. I. *Proc. Roy. Soc. (A)*, **275**, 23–32.

Jaswon, M. A. and Bhargava, R. D. (1961). Two-dimensional elastic inclusion problems. *Proc. Camb. Phil. Soc.* **57** (3), 669–680.

Jaswon, M. A. and Maiti, M. (1968). An integral equation formulation of plate bending problems. *J. Engng. Math.* **2** (1), 83–93.

Jaswon, M. A., Maiti, M. and Symm, G. T. (1967). Numerical biharmonic analysis and some applications. *Int. J. Solids Structures*, **3**, 309–332.

Jaswon, M. A. and Ponter, A. R. S. (1963). An integral equation solution of the torsion problem. *Proc. Roy. Soc. (A)*, **273**, 237–246.

Jeans, J. H. (1923). "The Mathematical Theory of Electricity and Magnetism." Fourth edition. Cambridge University Press.

Jones, D. S. (1966). "Generalised Functions." McGraw-Hill, London.

Kantorovich, L. V. and Krylov, V. I. (1964). "Approximate Methods of Higher Analysis." Noordhoff, Groningen.

Kellogg, O. D. (1929). "Foundations of Potential Theory." Springer, Berlin.

Kermanidis, T. (1976). Kupradze's functional equation for the torsion problem of prismatic bars—Part 1. *Computer Methods Appl. Mech. Engng.* **7**, 39–46.

Kermanidis, T. (1976). Kupradze's functional equation for the torsion problem of prismatic bars—Part 2. *Computer Methods Appl. Mech. Engng.* **7**, 249–259.

Knops, R. J. and Payne, L. E. (1971). "Uniqueness Theorems in Linear Elasticity." Springer-Verlag, Berlin.

Kupradze, V. D. (1965). "Potential Methods in the Theory of Elasticity." Israel Program for Scientific Translations, Jerusalem.

Kupradze, V. D. and Aleksidze, M. A. (1964). The method of functional equations for the approximate solution of certain boundary value problems. *USSR Comp. Maths. Math. Phys.* **4** (4), 82–126.

Lachat, J. C. and Watson, J. O. (1975). A second generation integral equation program for three-dimensional elastic analysis. *In* "Boundary-Integral Equation Method: Computational Applications in Applied Mechanics," (T. A. Cruse and F. J. Rizzo, Eds.), ASME, New York.

Lachat, J. C. and Watson, J. O. (1976). Effective numerical treatment of boundary integral equations: a formulation for three-dimensional elastostatics. *Int. J. Num. Methods Engng.* **10**, 991–1005.

Lamb, H. (1945). "Hydrodynamics." Sixth edition. Cambridge University Press.

Landkof, N. S. (1972). "Foundations of Modern Potential Theory." Springer-Verlag, Berlin.

Lardner, R. W. (1972). Dislocation layers and boundary-value problems of plane elasticity. *Quart. J. Mech. Appl. Math.* **25** (1), 45–61.

Laskar, S. K. (1971). "Solutions of Certain Boundary Integral Equations in Potential Theory." Ph.D. Thesis, The City University, London.

Laskar, S. K. (1974). "A Numerical Approach to Calculate Capacity of Conductors." Unpublished report, Department of Mathematics, Assam Engineering College, Gauhati.

Laura, P. A. (1965). "Conformal Mapping of a Class of Doubly Connected Regions." NASA Tech. Rep. No. 8, The Catholic University of America, Washington.

Lebesgue, H. (1913). Sur des cas d'impossibilité du problème de Dirichlet ordinaire. *C.R. Soc. Math. France,* **41**, 17.

Lebesgue, H. (1924). Conditions de régularité, conditions d'irrégularité, conditions d'impossibilité dans le problème de Dirichlet. *C.R. Acad. Sci. Paris,* **178**, 352–354.

Lehman, R. S. (1959). Developments at an analytic corner of solutions of elliptic partial differential equations. *J. Math. Mech.* **8**, 727–760.

Lighthill, J. (1976). Flagellar hydrodynamics—The John von Neumann Lecture, 1975. *SIAM Rev.* **18** (2), 161–230.

Lighthill, M. J. (1958). "Fourier Analysis and Generalised Functions." Cambridge University Press.

Lo, C. C. and Niedenfuhr, F. W. (1970). Singular integral equation solution for torsion. *J. Engng. Mech. Div., Proc. ASCE,* **96**, 535–542.

Love, A. E. H. (1927). "A Treatise on the Mathematical Theory of Elasticity." Fourth edition. Cambridge University Press.

Lynn, M. S. and Timlake, W. P. (1968a). The numerical solution of singular integral equations of potential theory. *Numerische Mathematik,* **11**, 77–98.

Lynn, M. S. and Timlake, W. P. (1968b). The use of multiple deflations in the numerical solution of singular systems of equations with applications to potential theory. *SIAM J. Numer. Anal.* **5**, 303–322.

Lynn, M. S. and Timlake, W. P. (1970). On the eigenvalues and eigenvectors of the integral equation formulation of the multi-interface Neumann problem. *J. Inst. Maths. Applics.* **6**, 391–399.

Maiti, M. and Chakrabarty, S. K. (1974). Integral equation solutions for simply supported polygonal plates. *Int. J. Engng. Sci.* **12** (10), 793–806.

Maiti, M. and Makan, G. (1973). Somigliana's method applied to elastic inclusions and dislocations. *J. Elasticity*, **3** (1), 45–49.

Maiti, M., Miller, G. F. and Symm, G. T. (1969). "Thermal stresses in polygonal plates." NPL DNAM Report 83.

Massonnet, C. E. (1965). Numerical use of integral procedures. *In* "Stress Analysis," (O. C. Zienkiewicz and G. S. Holister, Eds.). Wiley, New York.

Mautz, J. R. and Harrington, R. F. (1970). Computation of rotationally symmetric Laplacian potentials. *Proc. IEE*, **117** (4), 850–852.

McDonald, B. H., Friedman, M., Decreton, M. and Wexler, A. (1973). Integral finite-element approach for solving the Laplace equation. *Electronics Letters*, **9** (11), 242–244.

McDonald, B. H., Friedman, M. and Wexler, A. (1974). Variational solution of integral equations. *IEEE Trans. Microwave Theory and Techniques*, **22** (3), 237–248.

Mendelson, A. and Albers, L. U. (1975). Application of boundary integral equations to elastoplastic problems. *In* "Boundary-Integral Equation Method: Computational Applications in Applied Mechanics." (T. A. Cruse and F. J. Rizzo, Eds.). ASME, New York.

Mikhlin, S. G. (1957). "Integral Equations." Pergamon, London.

Mikhlin, S. G. (1965). "Multidimensional Singular Integrals and Integral Equations." Pergamon, London.

Mikhlin, S. G. (1970). "Mathematical Physics, an Advanced Course." North-Holland, Amsterdam.

Miller, G. F. (1974). Fredholm equations of the first kind. *In* "Numerical Solution of Integral Equations," (L. M. Delves and J. Walsh, Eds.). Clarendon, Oxford.

Mindlin, R. D. (1936). Notes on the Galerkin and Papkovich stress functions. *Bull. Amer. Math. Soc.* **42**, 373–379.

Morley, L. S. D. (1973). Finite element solution of boundary-value problems with non-removable singularities. *Phil. Trans. Roy. Soc.* (*A*), **275**, 463–488.

Motz, H. (1946). The treatment of singularities of partial differential equations by relaxation methods. *Quart. Appl. Math.* **4**, 371–377.

Muskhelishvili, N. I. (1953a). "Singular Integral Equations." Noordhoff, Groningen.

Muskhelishvili, N. I. (1953b). "Some Basic Problems of the Mathematical Theory of Elasticity." Noordhoff, Groningen.

Nabarro, F. R. N. (1967). "Theory of Crystal Dislocations." Clarendon, Oxford.

National Physical Laboratory. (1961). "Modern Computing Methods." Second edition. H.M.S.O., London.

Neuber, H. (1934). Ein neuer Ansatz zür losung räumlicher Probleme der Elastizitätstheorie. *J. Appl. Math. Mech.* (*ZAMM*), **14**, 203–212.

Noble, B. (1960). The numerical solution of the singular integral equation for the charge distribution on a flat rectangular lamina. *In* "Differential and Integral Equations, PICC Symposium, Rome." Birkhäuser Verlag, Basel.

Noble, B. (1971). Some applications of the numerical solution of integral equations to boundary value problems. *In* "Conference on Applications of Numerical Analysis," (J. Ll. Morris, Ed.). Springer-Verlag, Berlin.

Opfer, G. (1967). "Untere, beliebig verbesserbare Schranken für den Modul eines zweifach zusammenhängenden Gebietes mit Hilfe von Differenzenverfahren." Dissertation, Hamburg.

Palit, S. S. (1976). "On the Integral Equation Methods of Continuum Mechanics." Ph.D. Thesis, University of Jadavpur, Calcutta.

Papamichael, N. and Symm, G. T. (1975). Numerical techniques for two-dimensional Laplacian problems. *Computer Methods Appl. Mech. Engng.* **6**, 175–194.

Papamichael, N. and Whiteman, J. R. (1972). "A Numerical Conformal Transformation Method for Harmonic Mixed Boundary-Value Problems." Brunel University, Department of Mathematics Report No. TR/18.

Papamichael, N. and Whiteman, J. R. (1973). A numerical conformal transformation method for harmonic mixed boundary value problems in polygonal domains. *J. Appl. Math. Phys. (ZAMP)*, **24**, 304–316.

Papkovich, P. F. (1932). Solution générale des équations differentielles fondamentales d'élasticité exprimée par trois fonctions harmoniques. *C.R. Acad. Sci. Paris*, **195**, 513–515.

Peters, G. and Wilkinson, J. H. (1970). The least squares problem and pseudo-inverses. *Computer J.* **13** (3), 309–316.

Petrovsky, I. G. (1954). "Lectures on Partial Differential Equations." Interscience, New York.

Petrovsky, I. G. (1971). "Lectures on the Theory of Integral Equations." MIR, Moscow.

Phillips, H. B. (1934). Effect of surface discontinuity on the distribution of potential. *J. Maths. Phys.* **13**, 251–267.

Pitfield, R. A. and Symm, G. T. (1974). "Solution of Laplace's Equation in Two Dimensions." NPL Algorithms Library Document No. D3/01/0/Algol60/1/74.

Pogorzelski, W. (1966). "Integral Equations and their Applications." Pergamon, Oxford.

Ponter, A. R. S. (1966a). On plastic torsion. *Int. J. Mech. Sci.* **8**, 227–235.

Ponter, A. R. S. (1966b). An integral equation solution of the inhomogeneous torsion problem. *SIAM J. Appl. Math.* **14** (4), 819–830.

Priyakumari, K. H. (1977). Some Problems in Classical Hydrodynamics. Ph.D. Thesis, Indian Institute of Technology, Madras.

Protter, M. H. and Weinberger, H. F. (1967). "Maximum Principles in Differential Equations." Prentice-Hall, New Jersey.

Quinlan, P. M. (1964). The torsion of an irregular polygon. *Proc. Roy. Soc. (A)*, **282**, 208–227.

Reitan, D. K. and Higgins, T. J. (1951). Calculation of the capacitance of a cube by the method of subareas. *J. Appl. Phys.* **22**, 223–226.

Reitan, D. K. and Higgins, T. J. (1956). Accurate determination of the capacitance of a thin rectangular plate. *AIEE Trans.* **75** (1) 761–766.

Rim, K. and Henry, A. S. (1967). "An Integral Equation Method in Plane Elasticity." NASA Report, Department of Mechanics and Hydraulics, The University of Iowa, Iowa City.

Rim, K. and Henry, A. S. (1968). "Improvement of an Integral Equation Method in Plane Elasticity through Modification of Source Density Representation." NASA Report, Department of Mechanics and Hydraulics, The University of Iowa, Iowa City.

Rizzo, F. J. (1967). An integral equation approach to boundary value problems of classical elastostatics. *Quart. Appl. Math.* **25** (1) 83–95.

Rizzo, F. J. and Shippy, D. J. (1968). A formulation and solution procedure for the general non-homogeneous elastic inclusion problem. *Int. J. Solids Structures*, **4**, 1161–1179.

Savin, G. N. (1961). "Stress Concentration around Holes." Pergamon, London.

Segedin, C. M. and Brickell, D. G. A. (1968). Integral equation method for a corner plate. *J. Struct. Div., Proc. ASCE*, **94**, 41–52.

Shidfar, A. (1977). Private communication.

Shilov, G. E. (1965). "Mathematical Analysis." Pergamon, London.

Smirnov, V. I. (1964). "Integral Equations and Partial Differential Equations, a Course of Higher Mathematics," Vol. IV. Pergamon, London.

Smithies, F. (1958). "Integral Equations." Cambridge University Press.

Sommerfeld, A. (1964). "Mechanics of Deformable Bodies." Academic Press, New York and London.

Southwell, R. V. (1956). "Relaxation Methods in Theoretical Physics," Vol. II. Clarendon, Oxford.

Sterling, R. C. (1977). "The Use of Integral Equations of the First Kind in Potential Theory, Including a Consideration of the Possible Use of Eigenfunction Expansions." Research Memorandum No. 9, Department of Mathematics, The City University, London.

Symm, G. T. (1963). Integral equation methods in potential theory. II. *Proc. Roy. Soc.* (*A*), **275**, 33–46.

Symm, G. T. (1964). "Integral Equation Methods in Elasticity and Potential Theory." NPL Mathematics Report No. 51.

Symm, G. T. (1966). An integral equation method in conformal mapping. *Numerische Mathematik,* **9**, 250–258.

Symm, G. T. (1967). Numerical mapping of exterior domains. *Numerische Mathematik,* **10**, 437–445.

Symm, G. T. (1969a). Conformal mapping of doubly-connected domains. *Numerische Mathematik,* **13**, 448–457.

Symm, G. T. (1969b). External thermal resistance of buried cables and troughs. *Proc. IEE,* **116** (10), 1695–1698.

Symm, G. T. (1970). Numerical Solution of a Boundary-Value Problem." NPL DNAM Report 89.

Symm, G. T. (1973). "Treatment of Singularities in the Solution of Laplace's Equation by an Integral Equation Method." NPL Report NAC 31.

Symm, G. T. (1974). Computation of potential in a multiwire proportional counter of arbitrary cross-section. *Nucl. Instrum. Methods,* **118**, 605–607.

Symm, G. T. and Pitfield, R. A. (1974). "Solution of Laplace's Equation in Two Dimensions." NPL Report NAC 44.

Synge, J. L. (1957). "The Hypercircle in Mathematical Physics." Cambridge University Press.

Thwaites, B. (Ed.) (1960). "Incompressible Aerodynamics." Clarendon, Oxford.

Timoshenko, S. and Goodier, J. N. (1951). "Theory of Elasticity." Second edition. McGraw-Hill, New York.

Timoshenko, S. and Woinowsky-Kreiger, S. (1959). "Theory of Plates and Shells." Second edition. McGraw-Hill, New York.

Tottenham, H. (1970). A direct method for the solution of field problems. *Int. J. Num. Methods Engng.* **2**, 117–131.

Tropper, A. M. (1969). "Linear Algebra." Nelson, London.

Tsuji, M. (1959). "Potential Theory in Modern Function Theory." Maruzen, Tokyo.

Ugodčikov, A. G. (1955). Electromodelling of the conformal mapping of a circular cylinder onto a given doubly connected region. *Ukrain. Mat. Z.* **7**, 305–312.

van de Vooren, A. I. and Botta, E. F. F. (1972). Computation of potential flow around bodies. *In* "Numerical Methods in Fluid Dynamics," von Kármán Institute for Fluid Dynamics, Lecture Series 44.

Varga, R. S. (1962). "Matrix Iterative Analysis." Prentice-Hall, New Jersey.

Vlasov, V. K. and Bakushchinskii, A. B. (1963). The method of potentials and the numerical solution of Dirichlet's problem for the Laplace equation. *USSR Comp. Maths. Math. Phys.* **3** (3) 767–776.

Vogel, S. M. and Rizzo, F. J. (1973). An integral equation formulation of three dimensional anisotropic elastostatic boundary value problems. *J. Elasticity*, **3**, 203–216.

Voigt, W. (1928). "Lehrbuch der Kristallphysik mit Ausschluss der Kristalloptik." Teubner, Leipzig.

Voinov, V. V., Voinov, O. V. and Petrov, A. G. (1974). A method of computation of potential flow around a body of revolution in an incompressible fluid. *USSR Comp. Maths. Math. Phys.* **14**, 263–268.

Volterra, V. (1959). "Theory of Functionals and of Integral and Integro-Differential Equations." Dover, New York.

Wait, R. (1973). Finite-element-type solution of integral equations. *In* "International Computing Symposium 1973," (A. Günther, B. Levrat and H. Lipps, Eds). North-Holland, Amsterdam.

Wait, R. and Mitchell, A. R. (1971). Corner singularities in elliptic problems by finite element methods. *J. Comput. Phys.* **8**, 45–52.

Wang, S. T. and Blandford, G. E. (1976). Comparison of boundary integral equation and FE methods. *J. Struct. Div., Proc. ASCE*, **192**, 1941–1947.

Watson, J. O. (1972a). "The Analysis of Thick Shells with Holes, by Integral Representation of Displacement." Ph.D. Thesis, University of Southampton.

Watson, J. O. (1972b). The analysis of three dimensional problems of elasticity by integral representation of displacement. Paper presented at International Conference on Variational Methods in Engineering, Dept. Civil Engng., Southampton University, Sept. 25–29, 1972.

Weatherburn, C. E. (1962). "Advanced Vector Analysis." Bell, London.

Weinberger, H. F. (1965). "Partial Differential Equations." Blaisdell, New York.

Weinel, E. (1931). Die Integralgleichungen des ebenen Spannungszustandes und der Plattentheorie. *J. Appl. Math. Mech. (ZAMM)*, **11** (5), 349–360.

Whiteman, J. R. and Papamichael, N. (1971). "Numerical Solution of Two Dimensional Harmonic Boundary Problems Containing Singularities by Conformal Transformation Methods." Brunel University, Dept. of Mathematics Report No. TR/2.

Whiteman, J. R. and Papamichael, N. (1972). Treatment of harmonic mixed boundary problems by conformal transformation methods. *J. Appl. Math. Phys. (ZAMP)*, **23**, 655–664.

Wilkinson, J. H. (1965). "The Algebraic Eigenvalue Problem." Clarendon, Oxford.

Wilkinson, J. H. and Reinsch, C. (1971). "Linear Algebra, Handbook for Automatic Computation," Vol. II. Springer-Verlag, Berlin.

Willis, J. R. (1965). Dislocations and inclusions. *J. Mech. Phys. Solids*, **13**, 377–395.

Woods, L. C. (1953). The relaxation treatment of singular points in Poisson's equation. *Quart. J. Mech. Appl. Math.* **6**, 163–185.

Youngren, G. K. and Acrivos, A. (1975). Stokes flow past a particle of arbitrary shape: a numerical method of solution. *J. Fluid Mechs.* **69** (2), 377–403.

Zienkiewicz, O. C., Kelly, D. W. and Bettess, P. (1977). The coupling of the finite element method and boundary solution procedures. *Int. J. Num. Methods Engng.* **11**, 355–375.

Index